When Giants Shift: How Tiny Bacteria Adapt to a Warming World

Sakina

TABLE OF CONTENTS

Chapter 1

INTRODUCTION

Bacteria have lived on Earth for at least 3 billion years, surviving and evolving through many extinction events and changes in climate (1). While climate change is an existential threat to mankind, on the scale of bacterial life on Earth, a changing climate is par for the course. Yet, despite their obvious resilience, the bacterial response to a climate change has been overlooked, even in areas of life where bacteria are extremely important to humankind such as agriculture. Soil microbes are an integral part of soil health. They can help plants grow through nutrient acquisition, water retention, and protection from pathogens (2). Each of these roles will become even more important as the climate changes. In particular, regional precipitation will become more extreme, resulting in more drought and flooding events that will exacerbate land degradation.

One of the most well-studied plant growth-promoting bacteria are rhizobia. This group of bacteria fixes nitrogen (catalyzes reduction of N_2 gas to bioavailable ammonia) in symbioses with the legume (*Fabaceae*) family of plants. As a result of the symbiosis, the soil nitrogen content increases after legumes are grown in nitrogen depleted soil. For this reason, legumes are often used for crop rotations to increase soil nitrogen content for future plantings. In this way, rhizobia are an important key to maintaining soil health and even reversing land degradation. For rhizobia to be effective partners in agriculture into the future, we need to understand how they will cope with changes in climate (3).

The outer membrane is the first line of defense for gram-negative cells against environmental onslaughts such as those that bacteria may experience as the climate changes. The physical properties of the outer membrane influence its instant reaction to changing conditions. Two of the most important properties are stability and fluidity, but they often act in opposition to each other. High membrane stability that renders the membrane robust and less permeable is often achieved at the expense of membrane fluidity, which is critical for membrane protein function and lateral diffusion. The optimal biological membrane must balance these two goals of fluidity and stability to maintain a liquid ordered phase. The main way that bacteria modify the physical properties of the membrane is through modifying their membrane composition (4–6).

Rhizobia use many strategies to construct an outer membrane with optimal fluidity and stability. One of these strategies is to synthesize hopanoids, steroid-like lipids that are made by a significant fraction rhizobia (7). Compared to 10% of all sequenced bacteria, 33% of rhizobia genomes are predicted to contain *shc* (3, 8), which cyclizes squalene to make the simplest hopanoids which can undergo further modification (9–12). Hopanoids are known to make the membrane more rigid, increasing membrane stability (13–15). However, perhaps more importantly, diplopterol has been shown to condense the membrane while retaining lateral lipid diffusivity, a measure of fluidity, decoupling the relationship between fluidity and stability (14, 16).

In this work, I highlight the connection between the membrane properties of soil bacteria and the effects of climate change in the soil. In Chapter 2, I review our current understanding of the role of outer membrane components, such as hopanoids, in protecting rhizobia from their environment, specifically in the context of climate change. Chapter 3 is a detailed exploration of the importance of the extended class of hopanoids in the symbiosis between *Bradyrhizobium diazoefficiens* and the legume plant *Aeschynomene afraspera*, led by former postdoctoral scholar Brittany Belin. Previously, extended hopanoids were shown to be necessary in *B. diazoefficiens* under conditions experienced during symbiosis (17). However, Dr. Belin discovered that a *B. diazoefficiens* strain deficient in extended hopanoid biosynthesis ($\Delta hpnH$) was able to participate in a moderately successful symbiosis, with defects in symbiosis initiation. My contribution to this discovery was showing that the $\Delta hpnH$ strain was defective in motility, potentially explaining the discovered lag in the initiation phase. This result also supports the hypothesis that hopanoids play a role in maintaining proper membrane protein function, which I discuss further in Appendix 1. In Chapter 4, I partially resolve the paradox revealed in Chapter 3 by illustrating how physicochemical conditions, specifically osmolarity and divalent cation concentration, could partially compensate for the growth of the $\Delta hpnH$ strain at low pH. I discovered that hopanoids are conditionally essential in *B. diazoefficiens* and that extended hopanoids help the cell adjust to different environmental conditions. In Chapter 5, I conclude with remarks on the future of hopanoid research and advocate for a holistic approach to bioengineering or selecting improved rhizobia strains by considering their entire lifecycle and their experience of the effects of climate change.

References

1. Javaux EJ. 2019. Challenges in evidencing the earliest traces of life. Nature 572:451–460.
2. Miransari M. 2011. Soil microbes and plant fertilization. Appl Microbiol Biotechnol 92:875–885.
3. Tookmanian E, Belin BJ, Saenz J, Newman DK. 2021. The role of hopanoids in fortifying rhizobia against a changing climate. Environ Microbiol.
4. Ernst R, Ejsing CS, Antonny B. 2016. Homeoviscous adaptation and the regulation of membrane lipids. J Mol Biol 428:4776–4791.
5. Hazel JR, Williams EE. 1990. The role of alterations in membrane lipid composition in enabling physiological adaptation of organisms to their physical environment. Prog Lipid Res 29:167–227.
6. Chwastek G, Surma MA, Rizk S, Grosser D, Lavrynenko O, Rucińska M, Jambor H, Sáenz J. 2020. Principles of Membrane Adaptation Revealed through Environmentally Induced Bacterial Lipidome Remodeling. Cell Rep 32:108165.
7. Belin BJ, Busset N, Giraud E, Molinaro A, Silipo A, Newman DK. 2018. Hopanoid lipids: from membranes to plant-bacteria interactions. Nat Rev Microbiol 16:304–315.
8. Racolta S, Juhl PB, Sirim D, Pleiss J. 2012. The triterpene cyclase protein family: a systematic analysis. Proteins 80:2009–2019.
9. Siedenburg G, Jendrossek D. 2011. Squalene-hopene cyclases. Appl Environ Microbiol 77:3905–3915.
10. Seckler B, Poralla K. 1986. Characterization and partial purification of squalene-hopene cyclase from Bacillus acidocaldarius. Biochimica et Biophysica Acta (BBA) - General Subjects 881:356–363.
11. Bradley AS, Pearson A, Sáenz JP, Marx CJ. 2010. Adenosylhopane: The first intermediate in hopanoid side chain biosynthesis. Org Geochem 41:1075–1081.
12. Welander PV, Doughty DM, Wu CH, Mehay S, Summons RE, Newman DK. 2012. Identification and characterization of Rhodopseudomonas palustris TIE-1 hopanoid biosynthesis mutants. Geobiology 10:163–177.
13. Chen Z, Sato Y, Nakazawa I, Suzuki Y. 1995. Interactions between bacteriohopane-32,33,34,35-tetrol and liposomal membranes composed of dipalmitoylphosphatidylcholine. Biol Pharm Bull 18:477–480.
14. Mangiarotti A, Genovese DM, Naumann CA, Monti MR, Wilke N. 2019. Hopanoids, like sterols, modulate dynamics, compaction, phase segregation and permeability of membranes. Biochim Biophys Acta Biomembr 1861:183060.
15. Kannenberg E, Blume A, McElhaney RN, Poralla K. 1983. Monolayer and calorimetric studies of phosphatidylcholines containing branched-chain fatty acids and of their interactions with cholesterol and with a bacterial hopanoid in model membranes. Biochimica et Biophysica Acta (BBA) - Biomembranes 733:111–116.
16. Sáenz JP, Sezgin E, Schwille P, Simons K. 2012. Functional convergence of hopanoids and sterols in membrane ordering. Proc Natl Acad Sci USA 109:14236–14240.
17. Kulkarni G, Busset N, Molinaro A, Gargani D, Chaintreuil C, Silipo A, Giraud E, Newman DK. 2015. Specific hopanoid classes differentially affect free-living and symbiotic states of Bradyrhizobium diazoefficiens. MBio 6:e01251-15.

Chapter 2

THE ROLE OF HOPANOIDS IN FORTIFYING RHIZOBIA AGAINST A CHANGING CLIMATE

This chapter is adapted from: Tookmanian EM, Belin BJ, Sáenz JP, Newman DK. 2021. The role of hopanoids in fortifying rhizobia against a changing climate. Environ Microbiol 23:2906–2918. doi.org/10.1111/1462-2920.15594

Summary

Bacteria are a globally sustainable source of fixed nitrogen, which is essential for life and crucial for modern agriculture. Many nitrogen-fixing bacteria are agriculturally important, including the bacteria known as rhizobia, that participate in growth-promoting symbioses with legume plants throughout the world. To be effective symbionts, rhizobia must overcome multiple environmental challenges: from surviving in the soil, to transitioning to the plant environment, to maintaining high metabolic activity within root nodules. Climate change threatens to exacerbate these challenges, especially through fluctuations in soil water potential. Understanding how rhizobia cope with environmental stress is crucial for maintaining agricultural yields in the coming century. The bacterial outer membrane is the first line of defense against physical and chemical environmental stresses, and lipids play a crucial role in determining the robustness of the outer membrane. In particular, structural remodeling of Lipid A and sterol-analogues known as hopanoids are instrumental in stress acclimation. Here, we discuss how the unique outer membrane lipid composition of rhizobia may underpin their resilience in the face of increasing osmotic stress expected due to climate change, illustrating the importance of studying microbial membranes and highlighting potential avenues towards more sustainable soil additives.

Introduction

Though originally written with a marine context in mind, the famous lament, 'Water, water, everywhere; nor any drop to drink" from Samuel Taylor Coleridge's poem, "The Rime of the Ancient Mariner," could equally well describe the challenges climate change poses to the terrestrial realm. Chief among these paradoxical challenges are rising temperatures and changes in regional precipitation—with some areas predicted to experience extreme drought and others extreme flooding. According to a special report by the Intergovernmental Panel on Climate Change (IPCC), there is strong evidence that these impacts from climate change will accelerate desertification, land degradation, and food insecurity in the coming decades (1). Yet the IPCC states with high confidence that increasing food productivity has the potential to be one of the most effective responses to address climate change challenges to land (1).

Soil microbes can enhance food productivity by stimulating plant growth through a range of mechanisms, from acquisition of water and nutrients (*e.g.* nitrogen and phosphorus) to protecting plants from pathogens (2). Among the bacteria that can facilitate plant growth, the rhizobia are perhaps the best studied and most ubiquitous: a recent global atlas of the dominant bacteria found in soil placed the model rhizobium species *Bradyrhizobium diazoefficiens* at the top of its list (3). Rhizobia are a polyphyletic group that participate in symbioses with the legume (*Fabaceae*) family of plants, including the agriculturally important soybean, alfalfa, and peanut crops. Rhizobia form nodules within plant cells where they use a critical enzyme—nitrogenase—to catalyzes the conversion of dinitrogen gas to ammonium (fixed nitrogen) that is then provided to the plant and ultimately the soil.

The rhizobia-legume symbiosis is a sustainable alternative to chemical fertilization, which depends upon the major greenhouse gas-contributing Haber-Bosch process. For this reason, legumes are used in crop rotations to manage soil nitrogen content, and rhizobia have been used successfully at scale to fertilize large legume crops such as soybean in Brazil (4–6). However, field or seed inoculations, especially of nonindigenous strains, are often less successful than one might hope. The bacteria must be reapplied year after year because they fail to stably integrate into the soil community (7–10). These results highlight the importance of understanding how rhizobia survive the complex and ever-changing soil environment to improve the use of this sustainable symbiosis in the future.

Soil bacteria employ a variety of strategies to survive in the soil environment, including being metabolic jacks of all trades, producing antibiotics, and/or fortifying their membranes against environmental onslaughts. Given the challenges inherent to their dual lifestyles inhabiting both soil and plant roots, it is not surprising that rhizobia have evolved diverse ways to deal with environmental challenges, including adapting the protein and lipid components of their membranes (11–14). Due to their important and underappreciated roles in contributing to membrane robustness and fitness in the soil environment, we spotlight outer membrane lipids in this review. One fortifying lipid that rhizobia employ are hopanoids, a class of bacterial sterol-like lipids. Hopanoid biosynthesis is particularly overrepresented in the bradyrhizobia, and can confer tolerance to a wide range of stressors (from temperature, to pH, to antibiotics) (15).

As the climate changes, the continued success of these symbioses will depend upon the ability of rhizobia to survive physicochemical stresses arising from drastic changes in temperature and water potential (16). We highlight changes in water potential as a key stressor for three reasons. First, changing water potential has a large effect on soil microbial communities and is predicted to be the single greatest climate-related stressor that will impact plant-microbe interactions (17). Second, legumes are generally even more sensitive to osmotic stress than rhizobia, and these plants may rely on osmotic stress-tolerant soil bacteria to ameliorate salinity stress and boost water and nutrient uptake (18, 19). Finally, rhizobia experience water potential changes in two different phases of their life cycle: both in the bulk soil and when transitioning to life within the plant host.

Here we address how hopanoids may help rhizobia cope with the challenges of changing osmolarity. We begin with a brief summary of the life cycle of rhizobia, paying particular attention to the different microenvironmental stresses they are likely to encounter, especially as the climate changes. From this holistic perspective, we focus down several orders of magnitude to consider the outer membrane—both the first line of defense against environmental stresses and where many hopanoids reside. In the context of their interactions with other outer membrane molecules, we discuss the biophysical effects of hopanoids and how they may be particularly well-suited to protect bacteria against a range of osmotic stresses. We end with brief remarks about the potential for hopanoids to be used in the development of climate-resilient rhizobial soil additives.

The Life Cycle of Rhizobia

Before the rhizobia interact with the plant directly, they must survive the complex and far from static soil environment **(Figure 1)**. Soils are diverse habitats, with important chemical and physical parameters varying both regionally and locally. Bulk soils are classified into soil types based in part on the nature and amount of clay. They can differ profoundly with respect to their pH (acidic to alkaline), mineral and organic matter content, and particle size/density, among other parameters (20). Soil organic matter content and particle size further correlate with soil moisture retention, affecting the soil water potential over time (21). Moving beyond bulk soil, considerable variation in environmental parameters can also exist within soils at the scale that affects bacteria, including the water potential of pores at the microscale (22). It is also important to consider the effect of plants on the soil environment, especially in the rhizosphere—the realm of the soil in the vicinity of, and influenced by, plant roots. Plants secrete a wide range of organic molecules from their roots, including a large proportion of photosynthate comprising amino acids, organic acids and sugars that provide carbon sources that sustain the local microbial community (23). As legumes grow, they acidify the soil, both as a direct result of nitrogen fixation and as a strategy to solubilize phosphorous (23, 24).

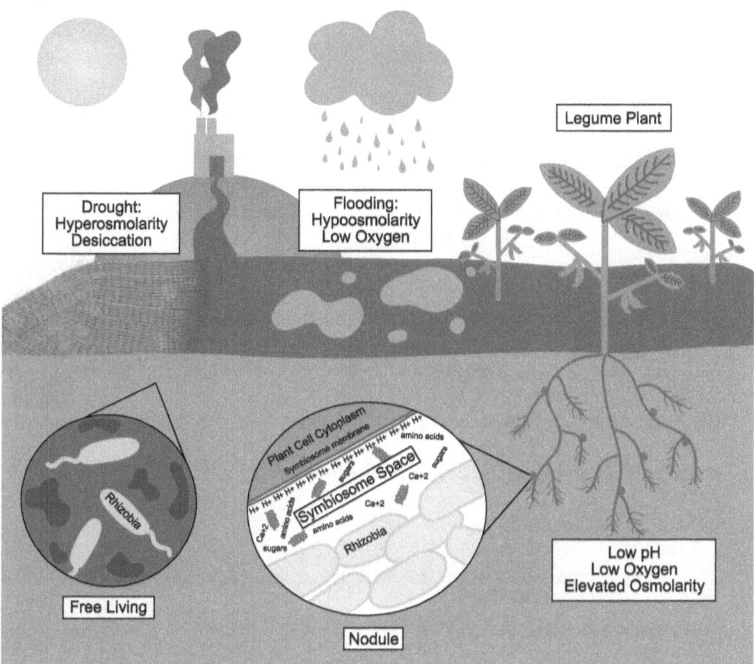

Figure 1. Dual Lifestyle of Rhizobia—from the Soil to the Plant. The free living stage in the soil or rhizosphere (the soil associated with and affected by plant roots) is shown on the left. Above the soil, precipitation changes (drought and flooding) as predicted to occur with climate change (indicated by the factory emitting fumes) are illustrated. The soil stresses the free living rhizobia may experience due to drought (hyperosmolarity and desiccation) or flooding (hypoosmolarity and low oxygen) are noted. To the right, a legume plant (e.g. *Glycine max,* soybean) is shown with its roots and circular root nodules. A close up of a root nodule is shown, illustrating the nodule stage. The symbiosome membrane is energized with protons in the symbiosome space, which also contains amino acids, sugars, calcium ions, and red helical leghemoglobin proteins. The stresses associated with the nodule environment (low pH, low oxygen, and elevated osmolarity) are noted.

Water potential is one of the most important factors for rhizobia that changing climate is predicted to affect (17). Precipitation will become more extreme: depending on geography, precipitation will either decrease, causing droughts and water scarcity, or increase, causing more extreme flooding. These precipitation changes directly influence water potential. During a drought, the water potential will decrease as the soil dries out, while during a rain event, water potential will increase. Varying water potential will have downstream effects on local soil chemistry by, for example, triggering changes in soil pH due to chemical weathering of mineral grains or altering local oxygen concentrations due to water saturation or stimulation of microbial metabolic activities (25). Strong shifts in osmolarity can also be expected, as it stands to reason that changing water potential will impact the concentration of solutes in a microenvironment: desiccation should render residual water pockets hyperosmotic, whereas flooding will render them hypoosmotic. If we can understand how bacteria cope with hypo- and hyperosmotic conditions, we can potentially engineer or select strains of rhizobia that will survive in soil conditions influenced by climate change.

Once the rhizobia have successfully managed to colonize the soil environment, they must make the transition to their legume hosts (**Figure 1**). Specifically, to initiate the symbiosis, rhizobia and the legume host exchange chemical signals that allow the rhizobia to colonize the root surface and invade the root tissue. Within the root some bacteria are taken up into plant cells where they are surrounded by a plant-derived symbiosome membrane. A visible structure known as a root nodule then develops via plant developmental programs and bacterial proliferation within the plant cytoplasm.

Between the symbiosome membrane and the bacterial outer membrane is the symbiosome space, where plant and bacterial proteins and metabolites comingle. To fully understand what stresses the bacteria experience during symbiosis, we must define the chemical and physical nature of the transition and of the symbiosome space microenvironment. However, directly characterizing these environments has been a challenge due to the lack of tools that accurately report the parameters of interest—e.g. pH, osmolarity, ionic strength—at the scale that matters in living tissues. Our current understanding has relied heavily on indirect evidence, using the fitness of bacterial mutants to infer the stress experienced, especially with regards to the process of nodule initiation (26–28). Additionally, different plant hosts produce different proteins and metabolites in nodules that affect the internalized bacteria, so we must resist generalizations based on any one legume-rhizobium partnership (29–32). Comparative studies across diverse legume-rhizobia partnerships are therefore needed to find underlying trends that may be obscured by studying a single partnership (33). Here, we focus on the consensus of the limited direct evidence.

We know that, regardless of host, the nodule environment is optimized for the energy-intensive process of nitrogen fixation. The legume partner provides essential molecules to fuel high metabolic activity, including abundant carbon sources derived from photosynthesis. Plant hosts also maintain a constant, low nanomolar concentration of oxygen using an oxygen-binding protein, leghemoglobin, which protects the oxygen-sensitive nitrogenase enzymes from inactivation while delivering sufficient oxygen to sustain bacterial aerobic respiration (34, 35). To facilitate active transport of essential nutrients (*i.e.* sugars, fixed nitrogen, iron) between the bacteria and the plant cell, the plant pumps protons into the symbiosome space. This process acidifies this compartment, which may partially aid the high metabolic rate of the bacteria but also adds a considerable stress. The acidic nature of the symbiosome space has been directly confirmed in *Medicago sativa* (alfalfa), indicating a pH between 4.5 and 5 (36). Rhizobia enter a new metabolic state, limiting many housekeeping processes in order to fix nitrogen for the plant, altering their metabolite output (37, 38).

The overall osmolarity of the symbiosome has not been quantified directly. One metabolomics study on the composition of the symbiosome space of *Glycine max* (soybean) nodules roughly approximated the concentration of low molecular weight organic compounds in the symbiosome space to 180 mM (39). However, this value does not account for inorganic ions or proteins, suggesting that the true symbiosome osmolarity could be higher, perhaps even approaching the approximate osmolarity of the bacterial cytoplasm (300 mOsM from studies in *Escherichia coli*) (40). The hypothesis that the symbiosome space has elevated osmolarity is supported by indirect lines of evidence, starting with the presence of compatible solutes in nodules. Compatible solutes, such as trehalose, are synthesized by bacteria in response to elevated osmolarity to increase their internal osmolarity in a way that is compatible with protein function (14, 41). Compatible solute biosynthesis and transport mutant studies in rhizobia have shown that compatible solutes are necessary for efficient symbiosis, from initiation to nitrogen fixation, again pointing to an elevated osmolarity in the symbiosome space (26, 42, 43).

Collectively, the available data indicate a symbiosome space environment with low pH, low oxygen, elevated osmolarity, and a comingling of bacterial and plant proteins and metabolites to facilitate nutrient conversion and exchange. While the symbiosome space presents many challenges to the bacteria, it appears to be a relatively static environment to promote nitrogen fixation (though how the symbiosome environment changes due to changes in the external environment has not been thoroughly investigated). In contrast, nodule initiation and living in the soil include inevitable shifts in different chemical parameters, including osmolarity. For both lifestyles and especially as the climate changes, rhizobia must be able to survive in a range of osmotic environments as well as transitions to different levels of osmolarity.

Rhizobial Outer Membrane

What strategies do rhizobia employ to facilitate survival in changing environments? They can respond by regulating the synthesis or activity of proteins, but in the first few seconds, they must rely on the outer membrane—the critical delimiter between the cell and the environment. The outer membrane's immediate, passive reaction to the changing conditions depends on the membrane's physical properties—most importantly, its fluidity and mechanical stability. These properties are determined by the lipid composition. Stability is important to maintain robustness and low permeability, but high stability usually comes at the expense of fluidity, which is critical for the function of membrane proteins that must diffuse laterally within the bilayer. Therefore, the optimal biological membrane is poised between two main opposing features: fluidity and stability (**Figure 2**). This generates a biophysical blind spot for cells constructing their surface membranes, which must be both stable and fluid. We will briefly summarize what is known about how the primary lipid components of the rhizobial outer membrane contribute to its fortification and maintain a crucial balance between fluidity and mechanical stability.

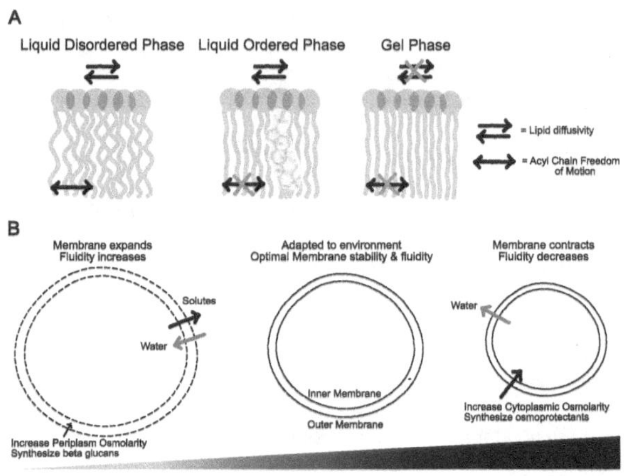

Figure 2. Physical Properties of Biological Membranes. (A) Three phases of lipid membranes are shown, starting with liquid disordered on the left, where both the acyl chains have freedom of movement and the lipid head groups can diffuse freely. In the liquid ordered phase in the middle, the lipid headgroups can still diffuse while the acyl chain movement is restricted. Finally, in the gel phase on the right, both the lipid headgroups and acyl chain movement is restricted. Diplopterol (a C_{30} hopanoid) is shown in the liquid ordered phase due to its ability to discourage the phase transitions into the liquid disordered or gel phase and encourage maintenance of the liquid ordered phase. Adapted from (44). (B) During changes in osmolarity, the cell undergoes a variety of changes which are illustrated here.

The gram-negative bacterial outer membrane is asymmetric with an inner leaflet made up of phospholipids and an outer leaflet containing lipopolysaccharides (LPS). LPS comprises lipid A (LA), consisting of a sugar backbone with various fatty acids tails that interact directly with other outer membrane phospholipids, and an outer region (the core oligosaccharide and O-antigen polysaccharide) that contributes to plant recognition of specific bacterial partners. Rhizobia synthesize structurally diverse LPS, but they all contain key differences in LA structure from the more well-studied *E. coli* (12) (**Figure 3**). Typical LA from *E. coli* consists of a glucosamine disaccharide backbone with two negatively charged phosphoryl groups on both ends of the disaccharide. The disaccharide is usually hexa-acylated with linear, saturated fatty acids 12-14 carbon atoms long. Almost all rhizobia LAs contain one striking difference: the presence of one or more very long chain fatty acids (VLCFA) attached to the LA backbone (accomplished in part by the enzyme LpxXL). For *Sinorhizobium* species, this is the only difference from the *E. coli* lipid A, changing from six regular fatty acids to four regular and one VLCFA. VLCFAs likely increase outer membrane stability and integrity due to increased hydrophobic interactions between the 28 carbon atom long VLCFA and the fatty acids of the outer membrane, possibly interacting with both outer membrane leaflets. Recently, it was shown that analogous asymmetric phospholipids with long and short acyl chains can enhance membrane stability in yeast (45). Interestingly, this LA modification is also found in intracellular pathogens like *Brucella* (46).

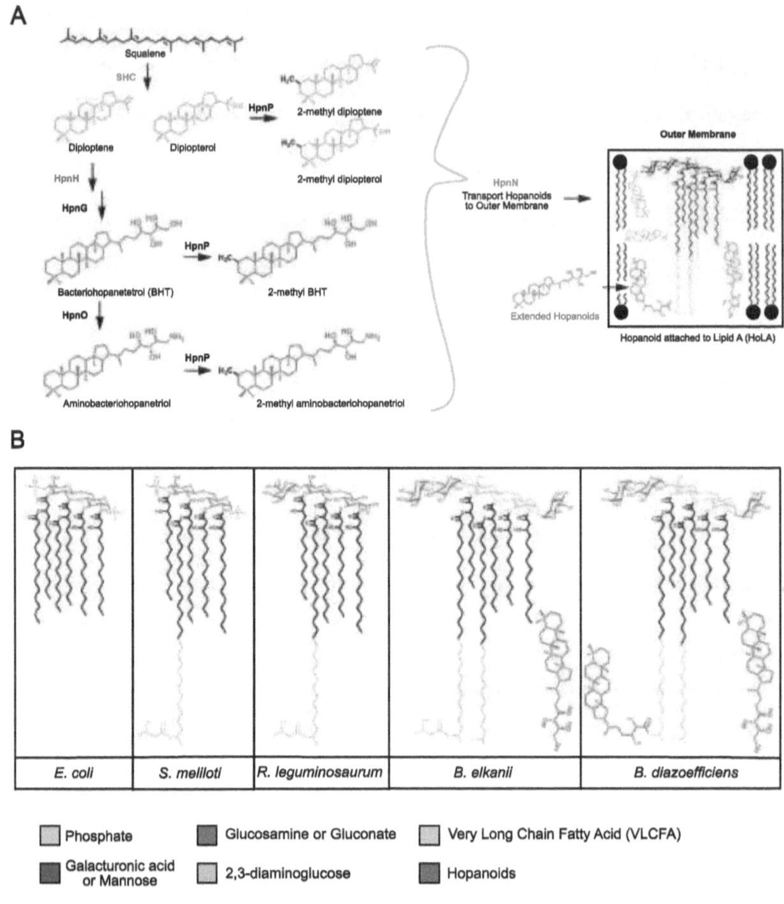

Figure 3. Hopanoid Biosynthesis and Outer Membrane Diversity in Rhizobia. (A) Hopanoid biosynthesis is shown starting with squalene (black) which is converted to the C_{30} hopanoids (magenta) by SHC. The C_{30} hopanoids are converted to different C_{35} hopanoids (green) by HpnH, HpnG, and HpnO. Next, these hopanoids can be 2-methylated (light blue) by HpnP. Hopanoids can be transferred to the outer membrane by the HpnN transporter. In the outer membrane, C_{35} hopanoids (likely BHT) can be attached to the VLCFA on Lipid A to make HoLA. (B) The outer membrane composition of different rhizobia is shown, with a comparison to the enteric bacterium, *E. coli* (left). Each structural change to lipid A, including the addition of hopanoids (C_{30}, magenta; C_{35}, green), is highlighted. The species' outer membrane compositions are shown from left to right in order of increasing resistance to polymyxin, a cationic antimicrobial peptide that targets the outer membrane (47).

The other major differences in LA structure are changes to the sugar backbone and substituents (**Figure 3**). Most rhizobia studied (except for *Sinorhizobium*) replace one or both phosphates on the ends of the disaccharide backbone, exchanging the negatively charged phosphate groups for protonated carboxylate groups or sugars with a neutral charge. This change may help when growing in a phosphorous poor environment. However, it also decreases electrostatic repulsion between LA molecules which might be important in the absence of bridging divalent cations like Mg^{+2} or Ca^{+2}, such as in aquatic conditions (*e.g.* flooded soil). The sugar backbone in some rhizobia (*Rhizobium* and *Sinorhizobium* strains) is the usual glucosamine disaccharide observed in *E. coli*. However, some studied *Meso-, Azo-,* and *Bradyrhizobium* strains alter the glucosamine so that the fatty acids are attached through amide linkages instead of the usual ester linkage (11). Unlike the ester linkages (O), the amide linkages (NH) may act as hydrogen bond donors, potentially further increasing cohesive lateral interactions between LA molecules (13). Each of these described modifications to LA increases cohesive interactions, likely increasing membrane order and stability compared to *E. coli*.

Hopanoids are another lipid class that may enhance rhizobial tolerance to environmental variability through their effects on the properties of the outer membrane. A significant fraction of rhizobia can make hopanoids. While about 10% of all sequenced bacteria have the genetic capacity to make hopanoids (48), as of writing 33% of the rhizobia genomes in the Integrated Microbial Genomes database are predicted to contain *shc*, the enzyme responsible for cyclizing squalene to make the simplest C_{30} hopanoids, diploptene or diplopterol (**Figure 3**), also known as "unextended" hopanoids (49, 50). Based on these computational predictions, within rhizobia, the ability to make hopanoids is mostly constrained to the *Bradyrhizobium* and *Burkolderia* clades, along with some *Methylobacteria* and three out of four *Sinorhizobium fredii* strains (**Figure 4**). However, the *S. fredii* strains do not appear to have the two hopanoid related genes that the other hopanoid-producing rhizobia contain: *hpnH* and *hpnN*. HpnH catalyzes the first committed step to make C_{35} hopanoids, also known as "extended" hopanoids, which include bacteriohopanetetrol (BHT) and bacterioaminotriol (51, 52) while HpnN transports hopanoids to the outer membrane (**Figure 3**) (53). Though methylation at the C-2 position of hopanoids was found to increase rigidity in native bacterial membranes (54), we did not determine the phylogenetic distribution of the C-2 methylase (HpnP) in our analysis because HpnP is not in the Pfam database.

In two species closely related to rhizobia that have the HpnN hopanoid transporter, hopanoids are preferentially trafficked to the outer membrane rather than remaining in the inner membrane, indicating hopanoids' importance to outer membrane function (44, 55). In *S. fredii*, the function of any C_{30} hopanoids synthesized is likely limited to the inner membrane due to the absence of *hpnN*. An additional subset of *Bradyrhizobia* make a LA variant where an extended hopanoid is attached to a VLCFA on LA. This variant is called hopanoid-attached LA (HoLA) (56–58), and HpnH is required for its synthesis (59). Five *Bradyrhizobia* species have been directly verified to produce HoLA: *B.* BTAi1, *B. diazoefficiens*, *B. yuanmingense*, *B.* sp. (Lupinus), and *B.* ORS278 (56–58). The enzyme responsible for this modification is not known, preventing the use of a computational approach to discover other species that can make HoLA.

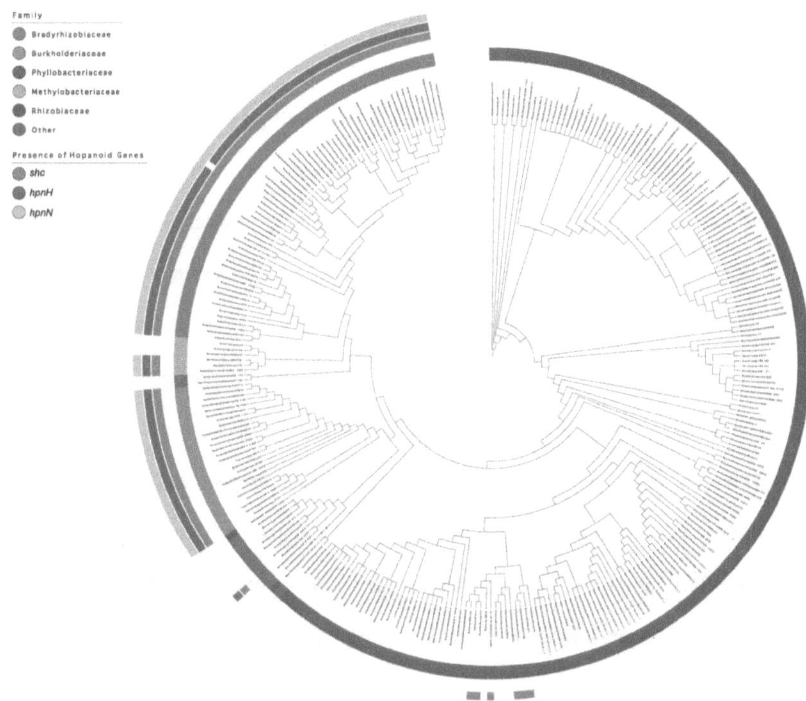

Figure 4. Phylogenetic Tree of Hopanoid Production in Rhizobia. Tree was built from 306 small subunit rRNA sequences from the Integrated Microbial Genomes (IMG) database using Silva Alignment, Classification, and Tree Service (ACT) which uses the FastTree program (60) and is rooted for display using the Interactive Tree of Life program (iTOL) (61). Rings identify the family of each species as well as whether their genomes contain hopanoid related genes from the ability to make any hopanoids (*shc* homolog, light pink) and/or C$_{35}$ hopanoids (*hpnH* homolog, green) and the ability to transport them to the outer membrane (*hpnN* homolog, grey) based on homology data from IMG.

The diversity of hopanoid structures that can be synthesized by bacteria indicate that different hopanoids are optimized for different functions in the cell, yet what these functions are remain to be determined. The most obvious difference occurs based on the presence or absence of a hydrophilic group. "Hydrophobic" hopanoids, like diploptene, that do not contain a hydrophilic group, are computationally predicted to localize at the midline of the bilayer, while hopanoids containing a hydrophilic group maintain an upright position with the hydrophilic group interacting with phospholipid head groups (62, 63). The "hydrophobic" hopanoids likely extrude water from the membrane and may be more important for membrane permeability than fluidity or stability. However, there are differences within the "hydrophilic" hopanoids as well. Both diplopterol and BHT have been shown to condense and order membranes containing saturated phospholipids (64–66) thus reducing membrane permeability (65, 67, 68). However, unlike cholesterol, diplopterol does not have any ordering or condensing effect on unsaturated lipids (44, 69). However, both the class of 2-methylated hydrophilic hopanoids and extended hopanoids are capable of ordering unsaturated phosphatidylcholine (54). These results show that modifications to the basic hopanoid ring structure can change its function and suggest that multiple hopanoid species collectively fulfill a comparable range of lipid ordering as cholesterol.

Importantly, part of what makes cholesterol so invaluable to eukaryotes is its ability to maintain a measure of fluidity through lateral lipid mobility (70). Cholesterol decouples the relationship between fluidity and robustness by forming a liquid ordered phase. While BHT has yet to be tested for this capability, diplopterol has been shown to keep lipids from entering a gel phase and retaining lateral lipid diffusivity (65, 71) (**Figure 2**). Specifically, hopanoids have a strong interaction with LA, which is especially interesting when thinking about HoLA's potential role in the membrane. Interestingly, the interaction of sterols with sphingolipids is both structurally and thermodynamically analogous to the interaction of hopanoids with Lipid A, suggesting that eukaryotes and bacteria have converged on a similar molecular solution to constructing a fluid and stable surface membrane (44).

In vitro work has shown that HoLA measurably increases the rigidity of the membrane (56). Despite only including HoLA in the outer leaflet of liposomes, HoLA had a pronounced rigidifying effect on both the outer and inner leaflet of the liposome. The VLCFA spans the OM, connecting both leaflets, while the attached hopanoid specifically condenses the inner leaflet, possibly even aiding insertion of the VLCFA to the inner leaflet. Addition of free hopanoids to the outer leaflet of a liposome containing LA with no hopanoid attached only ordered the outer leaflet, indicating that hopanoids alone do not have a cross leaflet ordering effect. The same experiment with hopanoids included in the inner leaflet or both leaflets was not examined, so it is unclear if HoLA can accomplish greater ordering than free hopanoids alone. Indeed, hopanoids are known to modulate the order of LA, perhaps indicating that there could be a synergistic effect of HoLA and free hopanoids in the outer membrane (71). However, the total ordering across both leaflets by HoLA is impressive, indicating that HoLA deserves special attention for its role in the OM. While much remains to be learned about the roles of diverse structural variants of hopanoids, the fact remains that whether free or bound to LA, hopanoids appear to help maintain an optimal balance between membrane stability and fluidity.

The Outer Membrane vs. Osmotic Stress

The outer membrane is on the front lines of any environmental change, but changes in osmolarity are known to specifically affect the membrane (**Figure 2**). *In vitro* evidence suggests that during hyperosmotic shock, lipid vesicles rigidify as water leaves the vesicle and vesicle surface area decreases, compressing the membrane (72). On the other end of the spectrum, hypoosmotic shock causes water to rush into lipid vesicles and the vesicle to expand (67). It is speculated that membranes become more fluid during this initial hypoosmotic shock, perhaps due to the increase in surface area as the vesicle expands to accommodate more water (73). Some whole cell evidence seems to contradict the *in vitro* membrane vesicle data described above showing an increase in whole cell membrane fluidity in response to increased osmolarity. However, these experiments used glycerol and polyethylene glycol which also affect membrane hydration and looked at the fluidity over longer time intervals than an initial shock (74, 75). This evidence illustrates the complicated nature of the cellular and membrane response to osmotic stress. Indeed, beyond the initial shock, the long-term membrane effects and adaptations to osmotic shifts are not well understood.

However, whether during an initial shock or adaptation, maintaining an optimal membrane state is important for bacterial survival of osmotic stress. One way to deal with a change in membrane state is to adjust the membrane composition to compensate, called homeoviscous adaptation (e.g. desaturating fatty acids to increase membrane fluidity in response to low temperatures) (76, 77). Rhizobia certainly utilize homeoviscous adaptations, but it is unclear whether the membrane modifications discussed in this review are under precise regulation, unlike in other bacteria such as *Salmonella* (78–80). Currently, we understand that the modified LA and hopanoids are produced by rhizobia under "normal" laboratory conditions, but a systematic exploration of the regulation of different types of hopanoids in rhizobia—particularly extended hopanoids and HoLA—is still needed. Studies in an organism closely related to rhizobia, *Rhodopseudomonas palustris,* found that unextended hopanoids are made during a variety of growth conditions but their abundance increases during stationary phase (81) and C-2 methylated hopanoids are regulated by the general stress response pathway (82). Given that the lack of extended hopanoids make cells particularly vulnerable to various outer membrane stresses (discussed below) and during stationary phase (59), it is possible that extended hopanoids and HoLA are also enriched under conditions that trigger outer membrane stress. Regardless, we propose that by making hopanoids and modified LA under unstressed conditions, rhizobia preemptively fortify their membrane in a bid for passive survival during challenges such as osmotic stress. Diplopterol (and potentially other hydrophilic hopanoids, including BHT) could protect the outer membrane from stresses by increasing stability (as previously discussed), reducing the sensitivity of membrane viscosity to physicochemical perturbations, and by suppressing the gel-liquid phase transition (71) thereby preventing the membrane from experiencing potentially catastrophic gel-liquid phase separation during rapid changes in physicochemical parameters such as pH and osmotic pressure.

Additionally, restricting the movement of water and solutes is another potential protective strategy. The most well understood adaptation of rhizobia to osmotic stress is the synthesis of sugars to balance the osmotic potential, either producing compatible solutes like trehalose in the cytoplasm during hyperosmotic stress or beta glucans in the periplasm during hypoosmotic stress (14). For these compatible solutes to be effective, membrane permeability needs to be maintained. Hopanoids and the rhizobial modifications to LA that increase cohesion may accomplish this by extruding water from the membrane, increasing lipid packing, and thus decreasing membrane permeability. Indeed, BHT has been shown to decrease the extent of lipid vesicle expansion when exposed to hypoosmotic shock (67).

Do hopanoids help rhizobia adapt to the range of osmotic stresses that are relevant to their lifestyle? Mutant analysis has been used to approach this question with respect to hyperosmotic stress, yet comes with important caveats: (i) deletion of any biosynthetic step can lead to accumulation of precursors or induce lipidome remodeling to compensate (83) and (ii) membrane protein function is tuned to its native lipid environment, so removing a membrane component likely results in broad reduction of membrane protein function (84, 85). In these analysis, HoLA-deficient mutants have been created by eliminating either hopanoids (*B*. BTAi1 Δ*shc* and *B. diazoefficiens* Δ*hpnH*) or the addition of VLCFAs on LA (*B*. ORS278 Δ*lpxXL*) (56, 58, 59, 86); the specific loss of HoLA has not yet been examined because the enzyme that conjugates hopanoids to lipid A remains unknown. All of these HoLA-deficient strains were sensitive to stresses affecting the OM, including hyperosmolarity, in free-living, rich-media cultures. Interestingly, both the *B*. BTAi1 Δ*shc* strain and the *B. diazoefficiens* Δ*hpnH* strain had growth defects under normal growth conditions, indicating the importance of free hopanoids as well as HoLA for general survival in these strains (56, 59). To date, no studies have looked at whether hopanoids confer protection to bacteria against hypoosmostic conditions—a frontier for future research—but the hyperosmotic growth defects observed in free-living rich-media conditions are likely magnified in the environment. We suggest that both hopanoids and HoLA/LPS may be key for rhizobial survival and competition in soils.

These HoLA-deficient mutants also were defective in symbiosis with various *Aeschynomene* legume hosts. The symbioses between either a *B*. BTAi1 Δ*shc* strain or a

B. ORS278 Δ*lpxXL* strain with *A. evenia* were similarly defective when compared to WT, eliciting a greater number of nodules that were largely ineffective at fixing nitrogen and beginning to senesce (56, 58). The *B. diazoefficiens* Δ*hpnH* mutant also showed defects in symbiosis initiation with a different species, *A. afraspera* (86). However, unlike the first two studies, observing nodules over a longer time course revealed that despite a lower nitrogen fixation rate early in symbiosis, with time, the *B. diazoefficiens* Δ*hpnH* mutant caught up to the WT. If all of these symbioses are able to recover, it may be that plant sterols incorporate into the bacterial membranes in the root nodule or that HoLA is only critical for initiation. Alternatively, if this result is specific to the *B. diazoefficiens* Δ*hpnH* mutant, it may be that the C_{30} hopanoids are sufficient after initiation. Intriguingly, hopanoid content in *B. diazoefficiens* actually decreases in root nodules and when grown in root extract, seeming to corroborate the view that hopanoids may be more important under osmotic transitions or hypoosmotic stress experienced in the free-living state or in symbiosis initiation than the static hyperosmolarity experienced within the root nodule(87). For species that do not make HoLA, VLCFAs on LA are similarly important. Mutants unable to add VLCFAs to LA showed defects in coping with diverse outer membrane stresses like hyperosmolarity and delayed or defective nodule development (88–93). These results illustrate a repeated theme for mutants in outer membrane components of rhizobia— a marked increase in sensitivity to environmental stressors, especially those related to outer membrane barrier function, along with a decrease in symbiotic efficiency, especially in initiation.

Concluding Remarks

As our climate changes and we look toward the future, the promise rhizobia hold for sustainable agriculture is undeniable. Rhizobia can decrease reliance on commercially produced nitrogen fertilizers and ameliorate negative effects of water potential extremes on plant productivity. To harness the potential of rhizobia, one route forward is bioengineering: selecting, evolving, or building rhizobia with improved symbiotic efficiency. While attention towards this end has been aimed at identifying traits that improve symbiosis *per se*, we suggest that it is important to keep a holistic perspective: to improve the fitness of plant-protecting organisms in a changing climate, we must also pay attention to traits that aid survival in the soil. This recognition motivates efforts to improve our mechanistic understanding of rhizobial adaptations in the context of a peripatetic lifestyle and dynamic changes to the soil environment. In this context, changing water potential is one of the most important parameters to be considered—be it inside of the plant or within the soil—given its dominant impact on bacterial soil populations and plant productivity (17, 94, 95). Water potential can span a large range in the soil, and whether a region is predicted to become more arid or flooded due to climate change varies globally. Yet flooded conditions that produce high water potential and hypoosmolarity are especially understudied and deserve more attention, particularly in light of the more frequent flooding expected for important agricultural regions (such as those in the upper Midwest of the United States or in India) in the coming decades. Given that hopanoids fortify the rhizobial outer membrane over a range of environmental challenges, tuning hopanoids' properties and expression may offer an attractive route to improving the ability of these important organisms to help agriculturally important crops cope with the challenges of a changing climate.

References

1. IPCC. 2019. Climate Change and Land: an IPCC special report on climate change, desertification, land degradation, sustainable land management, food security, and greenhouse gas fluxes in terrestrial ecosystems. In Press.
2. Miransari M. 2011. Soil microbes and plant fertilization. Appl Microbiol Biotechnol 92:875–885.
3. Delgado-Baquerizo M, Oliverio AM, Brewer TE, Benavent-González A, Eldridge DJ, Bardgett RD, Maestre FT, Singh BK, Fierer N. 2018. A global atlas of the dominant bacteria found in soil. Science 359:320–325.
4. Reckling M, Hecker J-M, Bergkvist G, Watson CA, Zander P, Schläfke N, Stoddard FL, Eory V, Topp CFE, Maire J, Bachinger J. 2016. A cropping system assessment framework—Evaluating effects of introducing legumes into crop rotations. European Journal of Agronomy 76:186–197.
5. Bullock DG. 1992. Crop rotation. CRC Crit Rev Plant Sci 11:309–326.
6. Loureiro M de F, Kaschuk G, Alberton O, Hungria M. 2007. Soybean [Glycine max (L.) Merrill] rhizobial diversity in Brazilian oxisols under various soil, cropping, and inoculation managements. Biol Fertil Soils 43:665–674.
7. Zhang NN, Sun YM, Li L, Wang ET, Chen WX, Yuan HL. 2010. Effects of intercropping and Rhizobium inoculation on yield and rhizosphere bacterial community of faba bean (Vicia faba L.). Biol Fertil Soils 46:625–639.
8. Roughley RJ, Gemell LG, Thompson JA, Brockwell J. 1993. The number of Bradyrhizobium SP. (Lupinus) applied to seed and its effect on rhizosphere colonization, nodulation and yield of lupin. Soil Biol Biochem 25:1453–1458.
9. Corich V, Giacomini A, Vendramin E, Vian P, Carlot M, Concheri G, Polone E, Casella S, Nuti MP, Squartini A. 2007. Long term evaluation of field-released genetically modified rhizobia. Environ Biosafety Res 6:167–181.
10. O'Callaghan M. 2016. Microbial inoculation of seed for improved crop performance: issues and opportunities. Appl Microbiol Biotechnol 100:5729–5746.
11. Choma A, Komaniecka I, Zebracki K. 2017. Structure, biosynthesis and function of unusual lipids A from nodule-inducing and N2-fixing bacteria. Biochim Biophys Acta Mol Cell Biol Lipids 1862:196–209.
12. Carlson RW, Forsberg LS, Kannenberg EL. 2010. Lipopolysaccharides in Rhizobium-legume symbioses. Subcell Biochem 53:339–386.
13. Nikaido H. 2003. Molecular Basis of Bacterial Outer Membrane Permeability Revisited. Microbiol Mol Biol Rev 67:593–656.
14. Miller KJ, Wood JM. 1996. Osmoadaptation by rhizosphere bacteria. Annu Rev Microbiol 50:101–136.
15. Belin BJ, Busset N, Giraud E, Molinaro A, Silipo A, Newman DK. 2018. Hopanoid lipids: from membranes to plant-bacteria interactions. Nat Rev Microbiol 16:304–315.
16. Cavicchioli R, Ripple WJ, Timmis KN, Azam F, Bakken LR, Baylis M, Behrenfeld MJ, Boetius A, Boyd PW, Classen AT, Crowther TW, Danovaro R, Foreman CM, Huisman J, Hutchins DA, Jansson JK, Karl DM, Koskella B, Mark Welch DB, Martiny JBH, Moran MA, Orphan VJ, Reay DS, Remais JV, Rich VI, Singh BK, Stein LY, Stewart FJ, Sullivan MB, van Oppen MJH, Weaver SC, Webb EA,

Webster NS. 2019. Scientists' warning to humanity: microorganisms and climate change. Nat Rev Microbiol 17:569–586.

17. Kardol P, Cregger MA, Campany CE, Classen AT. 2010. Soil ecosystem functioning under climate change: plant species and community effects. Ecology 91:767–781.

18. Enebe MC, Babalola OO. 2018. The influence of plant growth-promoting rhizobacteria in plant tolerance to abiotic stress: a survival strategy. Appl Microbiol Biotechnol 102:7821–7835.

19. Ilangumaran G, Smith DL. 2017. Plant growth promoting rhizobacteria in amelioration of salinity stress: A systems biology perspective. Front Plant Sci 8:1768.

20. Soil Survey Staff. 1999. Soil Taxonomy: A basic system of soil classification for making and interpreting soil surveys , 2nd ed. Natural Resources Conservation Service, U.S. Department of Agriculture Handbook 436.

21. Tiessen H, Cuevas E, Chacon P. 1994. The role of soil organic matter in sustaining soil fertility. Nature 371:783–785.

22. Tecon R, Or D. 2017. Biophysical processes supporting the diversity of microbial life in soil. FEMS Microbiol Rev 41:599–623.

23. Sugiyama A, Yazaki K. 2012. Root Exudates of Legume Plants and Their Involvement in Interactions with Soil Microbes, p. 27–48. *In* Vivanco, JM, Baluška, F (eds.), Secretions and exudates in biological systems. Springer Berlin Heidelberg, Berlin, Heidelberg.

24. Bolan NS, Hedley MJ, White RE. 1991. Processes of soil acidification during nitrogen cycling with emphasis on legume based pastures. Plant Soil 134:53–63.

25. Keiluweit M, Wanzek T, Kleber M, Nico P, Fendorf S. 2017. Anaerobic microsites have an unaccounted role in soil carbon stabilization. Nat Commun 8:1771.

26. Sugawara M, Cytryn EJ, Sadowsky MJ. 2010. Functional role of Bradyrhizobium japonicum trehalose biosynthesis and metabolism genes during physiological stress and nodulation. Appl Environ Microbiol 76:1071–1081.

27. Ledermann R, Bartsch I, Müller B, Wülser J, Fischer H-M. 2018. A Functional General Stress Response of Bradyrhizobium diazoefficiens Is Required for Early Stages of Host Plant Infection. Mol Plant Microbe Interact 31:537–547.

28. Gourion B, Sulser S, Frunzke J, Francez-Charlot A, Stiefel P, Pessi G, Vorholt JA, Fischer H-M. 2009. The PhyR-sigma(EcfG) signalling cascade is involved in stress response and symbiotic efficiency in Bradyrhizobium japonicum. Mol Microbiol 73:291–305.

29. Koch M, Delmotte N, Rehrauer H, Vorholt JA, Pessi G, Hennecke H. 2010. Rhizobial adaptation to hosts, a new facet in the legume root-nodule symbiosis. Mol Plant Microbe Interact 23:784–790.

30. Lardi M, Murset V, Fischer H-M, Mesa S, Ahrens CH, Zamboni N, Pessi G. 2016. Metabolomic Profiling of Bradyrhizobium diazoefficiens-Induced Root Nodules Reveals Both Host Plant-Specific and Developmental Signatures. Int J Mol Sci 17.

31. Mergaert P. 2018. Role of antimicrobial peptides in controlling symbiotic bacterial populations. Nat Prod Rep 35:336–356.

32. Czernic P, Gully D, Cartieaux F, Moulin L, Guefrachi I, Patrel D, Pierre O, Fardoux J, Chaintreuil C, Nguyen P, Gressent F, Da Silva C, Poulain J, Wincker P,

Rofidal V, Hem S, Barrière Q, Arrighi J-F, Mergaert P, Giraud E. 2015. Convergent Evolution of Endosymbiont Differentiation in Dalbergioid and Inverted Repeat-Lacking Clade Legumes Mediated by Nodule-Specific Cysteine-Rich Peptides. Plant Physiol 169:1254–1265.

33. Emerich DW, Krishnan HB. 2014. Symbiosomes: temporary moonlighting organelles. Biochem J 460:1–11.

34. Hunt S. 1993. Gas Exchange of Legume Nodules and the Regulation of Nitrogenase Activity. Annu Rev Plant Physiol Plant Mol Biol 44:483–511.

35. Appleby CA. 1984. Leghemoglobin and Rhizobium Respiration. Annu Rev Plant Physiol 35:443–478.

36. Pierre O, Engler G, Hopkins J, Brau F, Boncompagni E, Hérouart D. 2013. Peribacteroid space acidification: a marker of mature bacteroid functioning in Medicago truncatula nodules. Plant Cell Environ 36:2059–2070.

37. Yang Y, Hu X-P, Ma B-G. 2017. Construction and simulation of the Bradyrhizobium diazoefficiens USDA110 metabolic network: a comparison between free-living and symbiotic states. Mol Biosyst 13:607–620.

38. Terpolilli JJ, Masakapalli SK, Karunakaran R, Webb IUC, Green R, Watmough NJ, Kruger NJ, Ratcliffe RG, Poole PS. 2016. Lipogenesis and Redox Balance in Nitrogen-Fixing Pea Bacteroids. J Bacteriol 198:2864–2875.

39. Tejima K, Arima Y, Yokoyama T, Sekimoto H. 2003. Composition of amino acids, organic acids, and sugars in the peribacteroid space of soybean root nodules. Soil Sci Plant Nutr 49:239–247.

40. Stock JB, Rauch B, Roseman S. 1977. Periplasmic Space in Salmonella typhimurium and Escherichia coli. J Biol Chem 252:7850–7861.

41. Welsh DT. 2000. Ecological significance of compatible solute accumulation by micro-organisms: from single cells to global climate. FEMS Microbiol Rev 24:263–290.

42. Boscari A, Van de Sype G, Le Rudulier D, Mandon K. 2006. Overexpression of BetS, a Sinorhizobium meliloti high-affinity betaine transporter, in bacteroids from Medicago sativa nodules sustains nitrogen fixation during early salt stress adaptation. Mol Plant Microbe Interact 19:896–903.

43. Ledermann R. 2017. Role of general stress response in trehalose biosynthesis for functional rhizobia-legume symbiosis. Doctoral dissertation, ETH Zurich.

44. Sáenz JP, Grosser D, Bradley AS, Lagny TJ, Lavrynenko O, Broda M, Simons K. 2015. Hopanoids as functional analogues of cholesterol in bacterial membranes. Proc Natl Acad Sci USA 112:11971–11976.

45. Smith P, Owen DM, Lorenz CD, Makarova M. 2020. Asymmetric phospholipids impart novel biophysical properties to lipid bilayers allowing environmental adaptation. BioRxiv.

46. Bhat UR, Carlson RW, Busch M, Mayer H. 1991. Distribution and phylogenetic significance of 27-hydroxy-octacosanoic acid in lipopolysaccharides from bacteria belonging to the alpha-2 subgroup of Proteobacteria. Int J Syst Bacteriol 41:213–217.

47. Komaniecka I, Zamłyńska K, Zan R, Staszczak M, Pawelec J, Seta I, Choma A. 2016. Rhizobium strains differ considerably in outer membrane permeability and polymyxin B resistance. Acta Biochim Pol 63:517–525.

48. Racolta S, Juhl PB, Sirim D, Pleiss J. 2012. The triterpene cyclase protein family: a systematic analysis. Proteins 80:2009–2019.

49. Siedenburg G, Jendrossek D. 2011. Squalene-hopene cyclases. Appl Environ Microbiol 77:3905–3915.

50. Seckler B, Poralla K. 1986. Characterization and partial purification of squalene-hopene cyclase from Bacillus acidocaldarius. Biochimica et Biophysica Acta (BBA) - General Subjects 881:356–363.

51. Bradley AS, Pearson A, Sáenz JP, Marx CJ. 2010. Adenosylhopane: The first intermediate in hopanoid side chain biosynthesis. Org Geochem 41:1075–1081.

52. Welander PV, Doughty DM, Wu CH, Mehay S, Summons RE, Newman DK. 2012. Identification and characterization of Rhodopseudomonas palustris TIE-1 hopanoid biosynthesis mutants. Geobiology 10:163–177.

53. Doughty DM, Coleman ML, Hunter RC, Sessions AL, Summons RE, Newman DK. 2011. The RND-family transporter, HpnN, is required for hopanoid localization to the outer membrane of Rhodopseudomonas palustris TIE-1. Proc Natl Acad Sci USA 108:E1045-51.

54. Wu C-H, Bialecka-Fornal M, Newman DK. 2015. Methylation at the C-2 position of hopanoids increases rigidity in native bacterial membranes. Elife 4.

55. Wu CH, Kong L, Bialecka-Fornal M, Park S, Thompson AL, Kulkarni G, Conway SJ, Newman DK. 2015. Quantitative hopanoid analysis enables robust pattern detection and comparison between laboratories. Geobiology 13:391–407.

56. Silipo A, Vitiello G, Gully D, Sturiale L, Chaintreuil C, Fardoux J, Gargani D, Lee H-I, Kulkarni G, Busset N, Marchetti R, Palmigiano A, Moll H, Engel R, Lanzetta R, Paduano L, Parrilli M, Chang W-S, Holst O, Newman DK, Garozzo D, D'Errico G, Giraud E, Molinaro A. 2014. Covalently linked hopanoid-lipid A improves outer-membrane resistance of a Bradyrhizobium symbiont of legumes. Nat Commun 5:5106.

57. Komaniecka I, Choma A, Mazur A, Duda KA, Lindner B, Schwudke D, Holst O. 2014. Occurrence of an unusual hopanoid-containing lipid A among lipopolysaccharides from Bradyrhizobium species. J Biol Chem 289:35644–35655.

58. Busset N, Di Lorenzo F, Palmigiano A, Sturiale L, Gressent F, Fardoux J, Gully D, Chaintreuil C, Molinaro A, Silipo A, Giraud E. 2017. The Very Long Chain Fatty Acid (C26:25OH) Linked to the Lipid A Is Important for the Fitness of the PhotosyntheticBradyrhizobiumStrain ORS278 and the Establishment of a Successful Symbiosis withAeschynomeneLegumes. Front Microbiol 8:1821.

59. Kulkarni G, Busset N, Molinaro A, Gargani D, Chaintreuil C, Silipo A, Giraud E, Newman DK. 2015. Specific hopanoid classes differentially affect free-living and symbiotic states of Bradyrhizobium diazoefficiens. MBio 6:e01251-15.

60. Pruesse E, Peplies J, Glöckner FO. 2012. SINA: accurate high-throughput multiple sequence alignment of ribosomal RNA genes. Bioinformatics 28:1823–1829.

61. Letunic I, Bork P. 2019. Interactive tree of life (iTOL) v4: recent updates and new developments. Nucleic Acids Res 47:W256–W259.

62. Poger D, Mark AE. 2013. The relative effect of sterols and hopanoids on lipid bilayers: when comparable is not identical. J Phys Chem B 117:16129–16140.

63. Caron B, Mark AE, Poger D. 2014. Some like it hot: the effect of sterols and hopanoids on lipid ordering at high temperature. J Phys Chem Lett 5:3953–3957.

64. Chen Z, Sato Y, Nakazawa I, Suzuki Y. 1995. Interactions between bacteriohopane-32,33,34,35-tetrol and liposomal membranes composed of dipalmitoylphosphatidylcholine. Biol Pharm Bull 18:477–480.

65. Mangiarotti A, Genovese DM, Naumann CA, Monti MR, Wilke N. 2019. Hopanoids, like sterols, modulate dynamics, compaction, phase segregation and permeability of membranes. Biochim Biophys Acta Biomembr 1861:183060.

66. Kannenberg E, Blume A, McElhaney RN, Poralla K. 1983. Monolayer and calorimetric studies of phosphatidylcholines containing branched-chain fatty acids and of their interactions with cholesterol and with a bacterial hopanoid in model membranes. Biochimica et Biophysica Acta (BBA) - Biomembranes 733:111–116.

67. Bisseret P, Wolff G, Albrecht AM, Tanaka T, Nakatani Y, Ourisson G. 1983. A direct study of the cohesion of lecithin bilayers: The effect of hopanoids and α,ω-dihydroxycarotenoids. Biochem Biophys Res Commun 110:320–324.

68. Kannenberg E, Blume A, Geckeler K, Poralla K. 1985. Properties of hopanoids and phosphatidylcholines containing ω-cyclohexane fatty acid in monolayer and liposome experiments. Biochimica et Biophysica Acta (BBA) - Biomembranes 814:179–185.

69. Sáenz JP, Waterbury JB, Eglinton TI, Summons RE. 2012. Hopanoids in marine cyanobacteria: probing their phylogenetic distribution and biological role. Geobiology 10:311–319.

70. Mouritsen OG, Zuckermann MJ. 2004. What's so special about cholesterol? Lipids 39:1101–1113.

71. Sáenz JP, Sezgin E, Schwille P, Simons K. 2012. Functional convergence of hopanoids and sterols in membrane ordering. Proc Natl Acad Sci USA 109:14236–14240.

72. Yamazaki M, Ohnishi S, Ito T. 1989. Osmoelastic coupling in biological structures: decrease in membrane fluidity and osmophobic association of phospholipid vesicles in response to osmotic stress. Biochemistry 28:3710–3715.

73. Los DA, Murata N. 2004. Membrane fluidity and its roles in the perception of environmental signals. Biochim Biophys Acta 1666:142–157.

74. Cesari AB, Paulucci NS, Biasutti MA, Morales GM, Dardanelli MS. 2018. Changes in the lipid composition of Bradyrhizobium cell envelope reveal a rapid response to water deficit involving lysophosphatidylethanolamine synthesis from phosphatidylethanolamine in outer membrane. Res Microbiol 169:303–312.

75. Laroche C, Beney L, Marechal PA, Gervais P. 2001. The effect of osmotic pressure on the membrane fluidity of Saccharomyces cerevisiae at different physiological temperatures. Appl Microbiol Biotechnol 56:249–254.

76. Chwastek G, Surma MA, Rizk S, Grosser D, Lavrynenko O, Rucińska M, Jambor H, Sáenz J. 2020. Principles of Membrane Adaptation Revealed through Environmentally Induced Bacterial Lipidome Remodeling. Cell Rep 32:108165.

77. Sinensky M. 1974. Homeoviscous adaptation--a homeostatic process that regulates the viscosity of membrane lipids in Escherichia coli. Proc Natl Acad Sci USA 71:522–525.

78. Garcia-del Portillo F, Foster JW, Maguire ME, Finlay BB. 1992. Characterization of the micro-environment of Salmonella typhimurium-containing vacuoles within MDCK epithelial cells. Mol Microbiol 6:3289–3297.

79. Bader MW, Navarre WW, Shiau W, Nikaido H, Frye JG, McClelland M, Fang FC, Miller SI. 2003. Regulation of Salmonella typhimurium virulence gene expression by cationic antimicrobial peptides. Mol Microbiol 50:219–230.

80. Murata T, Tseng W, Guina T, Miller SI, Nikaido H. 2007. PhoPQ-mediated regulation produces a more robust permeability barrier in the outer membrane of Salmonella enterica serovar typhimurium. J Bacteriol 189:7213–7222.

81. Rashby SE, Sessions AL, Summons RE, Newman DK. 2007. Biosynthesis of 2-methylbacteriohopanepolyols by an anoxygenic phototroph. Proc Natl Acad Sci USA 104:15099–15104.

82. Kulkarni G, Wu C-H, Newman DK. 2013. The general stress response factor EcfG regulates expression of the C-2 hopanoid methylase HpnP in Rhodopseudomonas palustris TIE-1. J Bacteriol 195:2490–2498.

83. Neubauer C, Dalleska NF, Cowley ES, Shikuma NJ, Wu CH, Sessions AL, Newman DK. 2015. Lipid remodeling in Rhodopseudomonas palustris TIE-1 upon loss of hopanoids and hopanoid methylation. Geobiology 13:443–453.

84. Amin DN, Hazelbauer GL. 2012. Influence of membrane lipid composition on a transmembrane bacterial chemoreceptor. J Biol Chem 287:41697–41705.

85. Rice AJ, Alvarez FJD, Davidson AL, Pinkett HW. 2014. Effects of lipid environment on the conformational changes of an ABC importer. Channels 8:327–333.

86. Belin BJ, Tookmanian EM, de Anda J, Wong GCL, Newman DK. 2019. Extended Hopanoid Loss Reduces Bacterial Motility and Surface Attachment and Leads to Heterogeneity in Root Nodule Growth Kinetics in a Bradyrhizobium-Aeschynomene Symbiosis. Mol Plant Microbe Interact 32:1415–1428.

87. Kannenberg EL, Härtner T, Perzl M, Schmitz S, Poralla K. 1999. Hopanoid lipid content of bradyrhizobium bacteria is dependent on culture conditions, p. 41–44. *In* Martínez, E, Hernández, G (eds.), Highlights of nitrogen fixation research. Springer US, Boston, MA.

88. Bourassa DV, Kannenberg EL, Sherrier DJ, Buhr RJ, Carlson RW. 2017. The Lipopolysaccharide Lipid A Long-Chain Fatty Acid Is Important for Rhizobium leguminosarum Growth and Stress Adaptation in Free-Living and Nodule Environments. Mol Plant Microbe Interact 30:161–175.

89. Vanderlinde EM, Muszyński A, Harrison JJ, Koval SF, Foreman DL, Ceri H, Kannenberg EL, Carlson RW, Yost CK. 2009. Rhizobium leguminosarum biovar viciae 3841, deficient in 27-hydroxyoctacosanoate-modified lipopolysaccharide, is impaired in desiccation tolerance, biofilm formation and motility. Microbiology (Reading, Engl) 155:3055–3069.

90. Vedam V, Kannenberg E, Datta A, Brown D, Haynes-Gann JG, Sherrier DJ, Carlson RW. 2006. The pea nodule environment restores the ability of a Rhizobium leguminosarum lipopolysaccharide acpXL mutant to add 27-hydroxyoctacosanoic acid to its lipid A. J Bacteriol 188:2126–2133.

91. Vedam V, Haynes JG, Kannenberg EL, Carlson RW, Sherrier DJ. 2004. A Rhizobium leguminosarum lipopolysaccharide lipid-A mutant induces nitrogen-fixing nodules with delayed and defective bacteroid formation. Mol Plant Microbe Interact 17:283–291.

92. Vedam V, Kannenberg EL, Haynes JG, Sherrier DJ, Datta A, Carlson RW. 2003. A

Rhizobium leguminosarum AcpXL mutant produces lipopolysaccharide lacking 27-hydroxyoctacosanoic acid. J Bacteriol 185:1841–1850.

93. Ferguson GP, Datta A, Carlson RW, Walker GC. 2005. Importance of unusually modified lipid A in Sinorhizobium stress resistance and legume symbiosis. Mol Microbiol 56:68–80.

94. Gelybó G, Tóth E, Farkas C, Horel Á, Kása I, Bakacsi Z. 2018. Potential impacts of climate change on soil properties. Agrokémia és Talajtan 67:121–141.

95. Karmakar K, Rana A, Rajwar A, Sahgal M, Johri BN. 2015. Legume-Rhizobia Symbiosis Under Stress, p. 241–258. *In* Arora, NK (ed.), Plant microbes symbiosis: applied facets. Springer India, New Delhi.

Chapter 3

EXTENDED HOPANOID LOSS REDUCES BACTERIAL MOTILITY AND SURFACE ATTACHMENT AND LEADS TO HETEROGENEITY IN ROOT NODULE GROWTH KINETICS IN A *BRADYRHIZOBIUM-AESCHYNOMENE SYMBIOSIS*

Introduction

Hopanoids are steroid-like lipids that support bacterial survival under stress (reviewed in Belin et al. 2018). They are synthesized by the squalene-hopene cyclase (*shc*) family of enzymes (Ochs et al. 1992; Syren et al. 2016), which generate the pentacyclic, C_{30} hopanoid core from squalene. In many organisms, the C_{30} hopanoids can be further modified, including methylation at the C-2 position via the enzyme HpnP (Welander et al. 2010) and addition of a ribose-derived side chain by the enzyme HpnH (Fig. 1a) (Welander et al. 2012). Side chain-containing hopanoids are known collectively as the C_{35} or "extended" hopanoids and commonly include molecules with aminotriol-, polyol-, and adenosyl- side-chain moieties (Schmerk et al. 2015). Organism-specific side chains have also been observed, including a hopanoid-lipid A conjugate known as HoLA (Silipo et al. 2014; Kulkarni et al. 2015; Komaniecka et al. 2014) that so far has only been found in *Bradyrhizobiaceae*.

It is thought that hopanoids primarily promote bacterial survival by rigidifying and decreasing the permeability of membranes (Saenz et al. 2015; Wu et al. 2015), providing a better barrier against external stress. Structurally distinct hopanoids have different capacities to alter the biophysical properties of membranes and can also differ in the degrees of stress resistance they confer (reviewed in Belin et al. 2018). In the *Bradyrhizobia* genus of legume symbionts, hopanoids promote growth of free-living cultures under acid, salt, detergent, antibiotic, and redox stresses (Silipo et al. 2014; Kulkarni et al. 2015), and we previously showed that these stress resistance phenotypes are largely mediated by the extended hopanoid class (Kulkarni et al. 2015).

We also analyzed an extended hopanoid-deficient mutant of *Bradyrhizobium diazoefficiens* USDA110 in symbiosis with two legumes: the native soybean host for this strain and *Aeschynomene afraspera,* the native host of the closely related photosynthetic *Bradyrhizobia. A. afraspera* is a flood-tolerant legume from tropical West Africa, where it has been used in rice intercropping systems (Somado et al. 2003) and to accelerate wound healing in traditional medicine (Swapna et al. 2011; Chifundera 2001; Caamal-Fuentes et al. 2015; Lei et al. 2018). We found that extended hopanoid-deficient mutants of *B. diazoefficiens* fixed less nitrogen per nodule in *A. afraspera* than wild type, while this strain did not appear to have a defect in its native soybean host. Microscopy analyses of a small sample of extended hopanoid mutant-infected *A. afraspera* nodules revealed several aberrant cytological phenotypes, including both nodules containing necrotic signatures, disorganized infection zones, and visible starch granule accumulation (Kulkarni et al. 2015).

These phenotypes are common signatures of poor symbiont performance, yet the lack of genetic tools for *A. afraspera*, the limited literature on this host's response to non-cooperators compared to model plants, and the low number of nodules examined made it difficult to determine the underlying cause. While it has been proposed that hopanoids may enable high rates of symbiotic nitrogen fixation in some hosts by limiting oxygen diffusion across cell membranes (Vilcheze et al. 1994; Parsons et al. 1987; Abeysekera et al. 1990), from our previous assays, we could not determine whether the poor symbiotic performance of extended hopanoid mutants reflects ineffective nitrogen fixation *per se*, or is simply a consequence of lower general stress resistance. Because we did not observe an extended hopanoid mutant phenotype in soybean, we instead suggested that the extended hopanoid mutant may not survive exposure to nodule cysteine-rich (NCR) peptides, which are synthesized by *A. afraspera* (Czernic et al. 2015) but absent in soybean.

Here, we sought to dissect further the symbiotic phenotypes of *B. diazoefficiens* extended hopanoid mutants in association with *A. afraspera*. We found that the lower nitrogen fixation of extended hopanoid mutants can be fully explained by a reduction in root nodule sizes and rhizobial occupancy, indicating that the underlying defect is unrelated to per-bacteroid nitrogen fixation levels. Using a novel single-nodule tracking approach to quantify *A. afraspera* nodule development, we uncovered both the baseline growth parameters for wild-type nodules and a surprising heterogeneity of extended hopanoid mutant developmental phenotypes. These results challenge the conclusions of our prior study (Kulkarni et al. 2015) and identify new, potentially hopanoid-dependent stages in the *B. diazoefficiens*-*A. afraspera* symbiosis for future mechanistic work. This work also provides a quantitative reference point for understanding the impact of symbiotically ineffective strains on *A. afraspera* nodule development.

Results

Loss of extended hopanoids results in reduced nodule size

Previously, we observed a symbiotic defect for an extended hopanoid-deficient (Δ*hpnH*) strain of *B. diazoefficiens* in association with *A. afraspera* (Kulkarni et al. 2015). To further validate this defect, we inoculated *A. afraspera* plants with Δ*hpnH* (lacking extended hopanoids), Δ*hpnP* (lacking 2-Me hopanoids), or wild-type *B. diazoefficiens*. At 24 days post-inoculation (dpi), plants inoculated with Δ*hpnH* were shorter than wild-type-inoculated plants, although both strains produced equivalent numbers of nodules (**Fig. 1b**). Δ*hpnH*-inoculated plants also exhibited a roughly 50% decrease in the rate of acetylene gas reduction compared to wild-type-inoculated plants at this time point (**Fig. 1c**). In contrast, the Δ*hpnP* mutant was similar to wild type (**Fig. 1b-c**). These results are consistent with our previous findings (Kulkarni et al. 2015).

To assess Δ*hpnH* viability within *A. afraspera* nodules, we performed morphological analyses of nodules using confocal fluorescent microscopy. Fifty-seven wild-type and 67 Δ*hpnH* nodule cross-sections were stained with a bacterial Live:Dead kit, consisting of the cell-permeable SYTO9 dye (staining all cells) and propidium iodide (PI) (staining only cells with a compromised membrane). We did not observe an increase in predominantly PI-stained nodules for Δ*hpnH* compared to wild type (**Figs. S1,S2**). Signatures of plant necrosis, which we previously associated with Δ*hpnH* when we observed a smaller number of nodules (Kulkarni et al. 2015), occurred prominently in only 1/67 Δ*hpnH* nodules examined (**Fig. S2**). However, we did find that many Δ*hpnH* nodules contained disorganized infection zones and that Δ*hpnH* bacteroids were less elongated than WT (**Fig. 1d**), as we reported previously (Kulkarni et al. 2015).

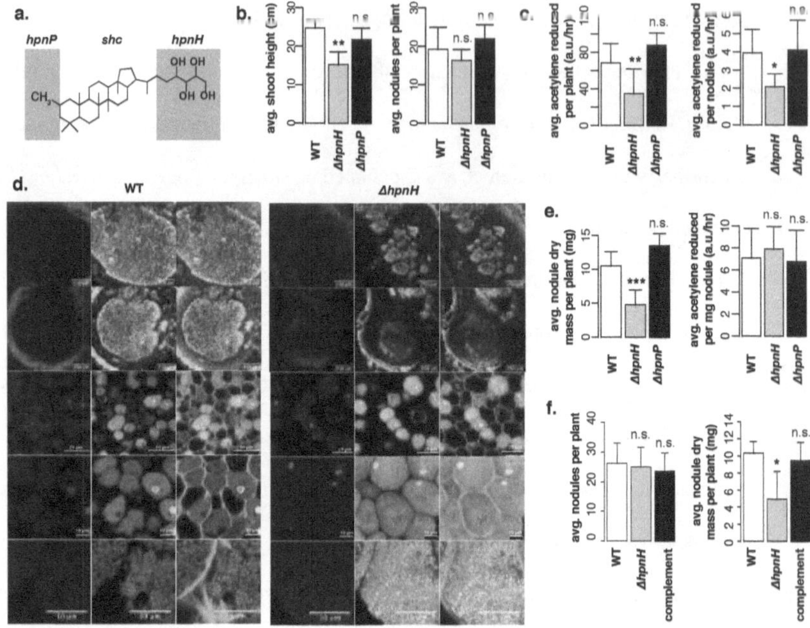

Figure 1. The nitrogen fixation defect of Δ*hpnH* results from a reduction in nodule sizes. (a) Chemical structure of the extended hopanoid 2-Methyl Bacteriohopanetetrol (2Me-BHT), consisting of a central pentacyclic core synthesized by the *shc* gene product, a C2 methylation site added by the product of *hpnP* (grey shading, left), and a tetrol group added by the *hpnH* product (grey shading, right). (b) Average shoot heights and nodules per plant at 24 dpi for *A. afraspera* plants inoculated with wild type, Δ*hpnH*, or Δ*hpnP B. diazoefficiens*. (c) Average acetylene reduction per plant and per nodule at 24 dpi for *A. asfrapera* plants inoculated with wild type, Δ*hpnH*, or Δ*hpnP*. (d) Representative confocal images of cross-sections of wild type- and Δ*hpnH*-infected nodules at 24 dpi illustrating plant cell walls (Calcofluor, cyan), live bacteria (SYTO9, yellow) and membrane-compromised bacteria and plant nuclei (propidium iodide, magenta). (e) Average nodule dry mass and acetylene reduction per nodule dry mass at 24 dpi for plants inoculated with wild type, Δ*hpnH*, or Δ*hpnP*. (f) Average number of nodules and nodule dry mass at 24 dpi for plants inoculated with wild type, Δ*hpnH* or a Δ*hpnH* complement strain. Data shown in (b), (c), (e) and (f) were collected from n = 8 plants, with error bars representing one standard deviation. Results of two-tailed t-tests between wild type and Δ*hpnH* or Δ*hpnP* are denoted as follows: n.s., p>0.05; *, p<0.01; **, p<0.001; ***, p<0.0001.

The most apparent phenotype of the Δ*hpnH* nodules was their relatively small size (**Fig. 1d; Figs. S1,S2**). We repeated acetylene reduction assays for wild-type- and Δ*hpnH*-inoculated plants and calculated the total nodule dry mass for each plant at 24 dpi. We found a decrease in the nodule dry mass per plant for Δ*hpnH*-inoculated plants that is sufficient to explain the decrease in acetylene reduction rates (**Fig. 1e**). This decrease in nodule dry mass can be fully rescued by integrating the *hpnH* gene at the endogenous *scoI* locus (**Fig. 1f**), suggesting that the lower nodule mass is due to *hpnH* loss specifically. This result rules out the possibility that nitrogenase functions ineffectively in the absence of extended hopanoids due to inactivation by oxygen, as has been suggested in *Frankia* (Vilcheze et al. 1994; Parsons et al. 1987; Abeysekera et al. 1990), as the per-mg nitrogen fixation rates are not affected by extended hopanoid loss.

*Δ*hpnH *nodules are more variable in size than wild-type nodules*

We measured acetylene reduction per plant across an extended 40 dpi period, and we observed that the differences in both acetylene reduction rates and nodule dry masses between wild type and Δ*hpnH* steadily decreased with time (**Fig. 2a-b**). By 40 dpi the overall symbiotic efficiencies of wild type and Δ*hpnH* per plant were indistinguishable, in terms of the plants' qualitative appearance (**Fig. 2c-d**) as well as their average shoot heights and acetylene reduction rates (**Fig. S3**). Total nodule counts per plant also did not differ between wild type and Δ*hpnH* at 40 dpi, indicating that the increase in total nodule mass reflects growing nodules rather than more frequent nodulation (**Fig. S3**).

We also measured the radii of individual nodules on ten plants for each strain at 40 dpi (**Fig. 2e-f**). Interestingly, although *average* nodule sizes did become similar between strains by this time point (0.73 vs. 0.88 mm average radii), their underlying distributions were markedly distinct. Wild-type nodule radii appear to form a roughly normal distribution, whereas the Δ*hpnH* nodule radius distribution is bimodal, consisting of a subpopulation of small nodules with small radii (<0.5 mm) that are rarely observed in wild type, as well as a second, larger subpopulation that has a similar median radius as wild type but is skewed towards larger radii (>1.5mm). These data demonstrate that the small-nodule phenotype of Δ*hpnH* persists throughout a 40 dpi time course, but is compensated by greater size heterogeneity, in which a handful of "mega" nodules offset smaller nodules over time.

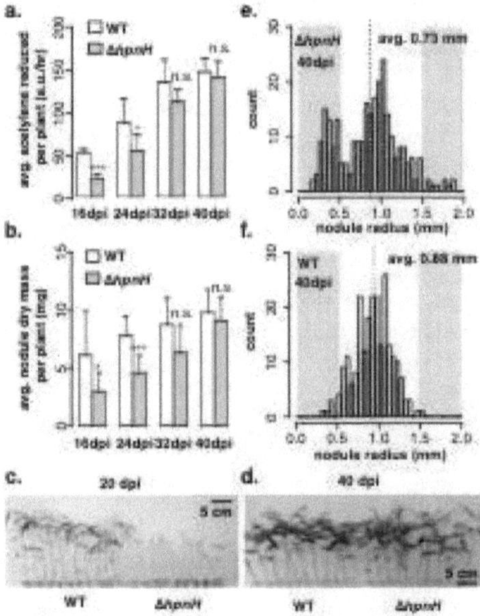

Figure 2. Smaller Δ*hpnH* nodules are offset by increased nodule size heterogeneity over time. (a) Average acetylene reduction per plant (n=4 plants per bar) and (b) average nodule dry mass per plant (n=8 plants per bar) for *A. asfrapera* inoculated with wild type or Δ*hpnH* over time. Error bars representing one standard deviation. Results of two-tailed t-tests between wild type and Δ*hpnH* are denoted as follows: n.s., p>0.05; *, p<0.05; ***, p<0.0001. (c-d) *A. afraspera* inoculated with wild type or Δ*hpnH* at (c) 20 dpi (left) and at (d) 40 dpi (right). (e-f) Distributions of nodule diameters at 40 dpi for *A. afraspera* inoculated with (e) Δ*hpnH* (right; n=268 nodules pooled from 10 plants) or (f) wild type (left; n=227 nodules pooled from 10 plants).

*Δ*hpnH *nodule size heterogeneity reflects variable nodule growth rates*

To better evaluate the possible origins of the $\Delta hpnH$ nodule size defect, we studied the kinetics of single nodule development. Beginning one week after inoculation, we collected images of entire plant roots every 3-5 days up to ~40 days post-inoculation (**Fig. S4,S5**). From these images, we identified nodules that were clearly visible (*e.g.* not obscured by lateral roots or more recently emerged nodules) in at least five time points (**Fig. 3a**) and measured their radii. We calculated nodule volumes by approximating nodules as spheres and plotted the volume of the tracked nodules over time. While we again observed that many $\Delta hpnH$ nodules were smaller at 40 dpi than any of the wild-type nodules, we also found that nodule growth was highly variable both within and between strains (**Fig. 3b-c**).

We developed a simple framework for quantifying nodule development, in which nodule growth is defined by the following variables: the time (t_i) of the initial intracellular infection event and the volume of the nascent nodule (V_i), equivalent to the volume of one infected *A. afraspera* cortical cell; the time (t_{min}) and volume (V_{min}) at which a clearly visible, spherical nodule has developed; the rate of growth of a nodule once it has become visible (dV/dt); and the time (t_{max}) and volume (V_{max}) of a nodule when its growth has stopped (**Fig. 3d**). To calculate these variables, we fit each nodule's growth over time to three different growth models: exponential, quadratic, and a generalized logistic (*e.g.* sigmoidal) equation commonly used for plant growth (Szparaga and Kocira 2018; Richards 1959) (see Methods for complete details). Sigmoidal models generally provided the best fit to the experimental data, so these models were used for growth parameter calculation (**Fig. 3e; Figs. S6, S7**).

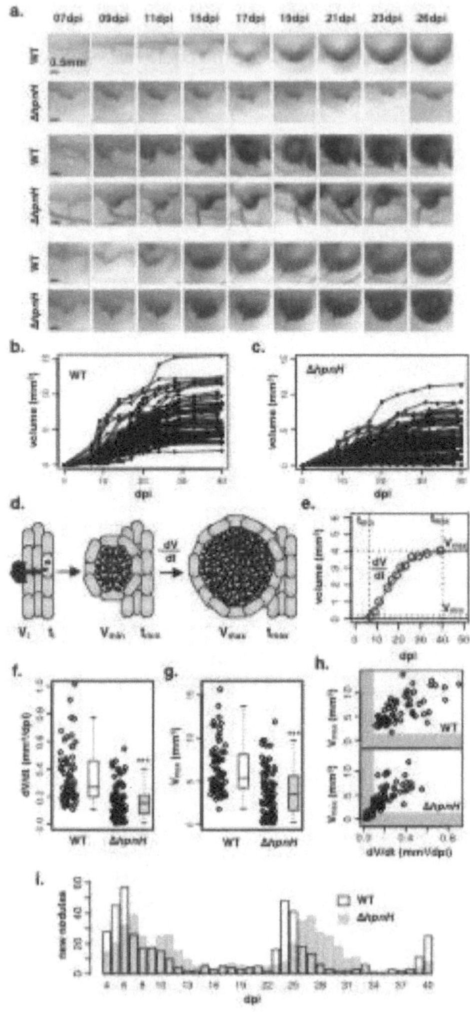

Figure 3. Nodules containing Δ*hpnH* emerge later and have more heterogeneous growth rates and final volumes than wild type. (a) Comparison of the development of selected wild type- and Δ*hpnH*-infected nodules over time. (b) Nodule growth plots for 74 wild-type-infected nodules tracked from 10 plants. (c) Nodule growth plots for 84 Δ*hpnH*-infected nodules tracked from 16 plants. (d) Schematic of nodule development in *A. afraspera*. From the left, bacteria (in blue) colonize and invade plant roots (green) and intracellularly infect a root cell (pink); the time of this initial intracellular infection is considered t_i and the nodule volume can be described as the volume of the single infected root cell, V_i. This infected cell proliferates to form a spherical nodule that is visible to the naked eye, at time t_{min} and volume V_{min}. The infected plant cells continue to proliferate at rate **dV/dt** until the nodule has fully matured at time t_{max} and volume V_{max}. (e) Fitted growth curve for a sample wild-type nodule illustrating the positions of t_{min}, V_{min}, **dV/dt**, t_{max}, and V_{max}. (f-g) Jitter and box plots of (f) **dV/dt** and (g) V_{max} values for all wild-type- and Δ*hpnH*-infected nodules. Results of KS-tests between wild-type and Δ*hpnH* nodules are denoted as follows: ***, $p < 10^{-6}$. (h) Scatter plots of **dV/dt** vs. V_{max} values for wild-type and Δ*hpnH* nodules. Values of **dV/dt** and V_{max} below what is observed in the wild-type dataset are highlighted in green. (i) Distributions of t_{min} values (as observed by eye) for nodules from wild-type- (white bars) or Δ*hpnH*- (grey bars) infected plants. N=457 wild-type nodules across 20 plants and 479 Δ*hpnH* nodules across 20 plants.

The growth rates of $\Delta hpnH$ nodules were lower on average than wild-type nodules (**Fig. 3f**), with roughly a third of tracked nodules exhibiting growth rates lower than observed for wild type (<0.1 mm^3/dpi). A similar fraction of nodules had smaller final volumes than wild type (**Fig. 3g**). We further found that the growth rate of a nodule and its maximum size are positively linearly correlated for both strains, with Pearson coefficients of ~0.64 ($p<10^{-9}$) for wild type and ~0.75 ($p<10^{-15}$) for $\Delta hpnH$, and that the subpopulation of nodules with lower-than-wild-type growth rates and small nodule sizes are the same (**Fig. 3h**). We interpret these data to suggest that host cell proliferation and/or host cell expansion is slower in a subset of nodules infected with $\Delta hpnH$, and that this largely accounts for the low final volume of these nodules.

We also noted that $\Delta hpnH$ nodule sizes at 40 dpi differed between these single-nodule volume measurements (**Fig. 3g**) and our previous 40 dpi end-point measurements of nodule radii (**Fig. 2e**), in that we did not observe larger-than-wild-type "mega" nodules in the single-nodule dataset. This discrepancy likely reflects the smaller sample size in our single-nodule tracking experiments (84 compared to 268 end-point nodules), and the low frequency of "mega" nodule formation. To verify this, we selected 10,000 random subsets of 84 nodules from the 268 $\Delta hpnH$ nodules shown in **Figure 2e**, converted the nodule radii to volumes, and found that there is no statistically significant difference ($p<0.05$) between a random subset of **Fig. 2e** and the $\Delta hpnH$ single-nodule tracking data in ~92% (9184/10000) of cases. Thus the differences in nodule size distributions in Figure 3g and Figure 2e are consistent with the sampling error.

We also calculated each nodule's window of maximum growth, defined as the time required for a nodule to increase from 10% to 90% of its final volume. Neither the time at which a nodule reaches 90% of its maximum volume, t_{max}, nor the window of maximum growth differs significantly between Δ*hpnH* and wild type (**Fig. S8a-b**). The window of maximum growth for each nodule is also uncorrelated with their final volume or growth rate, indicating that small nodules are not prematurely aborted; rather, their growth periods are similar to larger nodules (**Fig. S9a-d**).

To better understand the subpopulation of small, slow-growing Δ*hpnH* nodules, we isolated nodules with <0.5 mm radius, sectioned and stained them with SYTO9, PI, and Calcofluor, and imaged them with confocal microscopy. We found that while most small Δ*hpnH* nodules contained a single, continuous infection zone, a large fraction were un- or under-infected with bacteria, often exhibiting disorganized central infection zones (~37%; 28/75) (**Fig. 4a; Fig. S10**). Of the fully infected small Δ*hpnH* nodules, a subset contained primarily PI-stained, likely dead bacterial cells (~25%; 12/47) (**Fig. 4a; Fig. S10**). Similar proportions of under-infected nodules or nodules primarily occupied with membrane-compromised bacteria did not occur in larger Δ*hpnH* nodules harvested at the same time point, although fragmented infection zones were still common (**Fig. 4b; Fig. S11**). We also compared the subpopulation of small Δ*hpnH* nodules at 40 dpi to two wild-type nodule populations: similarly small nodules harvested at 10 and 25 dpi (**Fig S12; Fig. S13**), and nodules harvested at the same 40 dpi time point (**Fig. S14**). Again, we found that high proportions of under-infected nodules and membrane-compromised bacteria were unique to the Δ*hpnH* small-nodule subset.

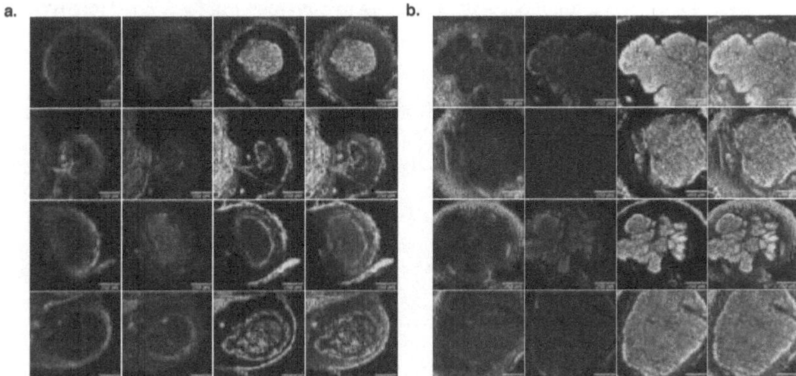

Figure 4. Small Δ*hpnH* nodules are under-infected compared to wild type. (a) Confocal sections of small (<0.5 mm radius) Δ*hpnH*-infected nodules harvested at 40 dpi. (b) Confocal sections of larger (>0.5 mm radius) Δ*hpnH*-infected nodules harvested at 40 dpi.

*Δ*hpnH *nodule emergence is delayed*

The "true" beginning of nodule formation is the time when the first *A. afraspera* cortical cell is infected, t_i (**Fig. 3d**). However, this initial infection event is not visible at the root surface, and it is difficult to extrapolate from sigmoidal models in which the growth curves approach the initial volume $V_i \sim 0$ mm^3 asymptotically. As a proxy for t_i, we defined three alternate t_{min} as the times at which nodules reached three arbitrarily small volumes: $V = 0.05$ mm^3, $V = 0.1$ mm^3, and $V = 0.2$ mm^3. When t_{min} is defined by $V = 0.05$ mm^3 or 0.1 mm^3, t_{min} could not be accurately calculated for all nodules, as the sigmoidal models sometimes predicted an impossible $t_{min} < 0$ (**Fig. S8c**). These nodule volumes are also too small to be seen on the root surface, and we had no experimental means to determine the accuracy of the calculations in this low-volume regime. When t_{min} is defined by $V = 0.2$ mm^3 (the smallest nodule volume that we could identify in our single-nodule tracking assays), there is a small but statistically significant increase for $\Delta hpnH$ relative to wild type (**Fig. S8c**).

To independently verify this delay in nodule emergence, we inspected the roots of 20 wild-type- and 20 $\Delta hpnH$-inoculated plants over 40 dpi and recorded the number of visible nodules per plant each day. We found a more even distribution of observed t_{min} for $\Delta hpnH$ relative to wild type, with a 1-3 day shift in the most frequent dpi. Surprisingly, we also found that the formation of new nodules is periodic, with a new "burst" of nodules emerging roughly every 18 days (**Fig. 3i**). This periodicity of nodule emergence appears to be similar between strains.

While the slight t_{min} delay for $\Delta hpnH$ is consistent with longer times required to initiate the symbiosis (*e.g.* root surface colonization, invasion of the root epidermis and cortex, and intracellular uptake), it is also possible that a delay in t_{min} simply reflects a lower rate of nodule growth immediately after the first intracellular infection. To address this, we compared the calculated value of t_{min} (defined by $V = 0.2$ mm^3) to the maximum growth rates and volumes for each nodule (**Fig. S9e-f**). We did not find that nodules with lower growth rates and final volumes than wild type were more likely to have a later t_{min}, supporting the interpretation that the delay in t_{min} of $\Delta hpnH$ could be due to a separate initiation defect. Interestingly, t_{min} is also not correlated with the period in which maximum nodule growth occurs, such that later-emerging nodules have similar growth period to nodules formed within a few dpi (**Fig. S9g-h**). This indicates that although nodule emergence is restricted to narrow, periodic windows (**Fig. 3i**), once a nodule has entered its maximum growth phase, its continued growth is comparatively unconstrained.

*Δ*hpnH *is delayed in a pre-endosymbiont stage*

We performed competition assays using a standard fluorescence labeling approach. We first generated Δ*hpnH* and wild-type strains expressing chromosomally-integrated fluorescent proteins, and we co-inoculated *A. afraspera* with different ratios of these two strains. As control experiments, we also co-inoculated each tagged strain with its untagged counterpart, in order to determine the effect of fluorescent protein overexpression on each strain's competitiveness. After 40 dpi we measured the size of nodules on plants inoculated with each strain combination and ratio and sectioned and fixed nodules for imaging. Although we expected each nodule to contain a clonal population of symbionts based on previous work (Bonaldi et al. 2011; Ledermann et al. 2015), the majority of nodules instead contained a mixture of both strains (**Fig. 5a**).

We quantified the relative abundance of each strain in each nodule by fluorescence imaging; in our control experiments, in which only one fluorophore-expressing strain was present, a DNA dye was used to label all bacteria. Both WT-YFP and Δ*hpnH*-mCherry were significantly out-competed by their corresponding untagged strains. In nodules with higher proportions of tagged strains, we observed lower bacterial DNA abundance and smaller nodule and/or infection zone sizes (**Fig. 5b-c; Figs. S16-S18**). Additionally, plants co-inoculated with untagged-Δ*hpnH* and Δ*hpnH*-mCherry were significantly shorter than plants inoculated with untagged-Δ*hpnH* only, suggesting Δ*hpnH*-mCherry is symbiotically defective (**Fig. S15**).

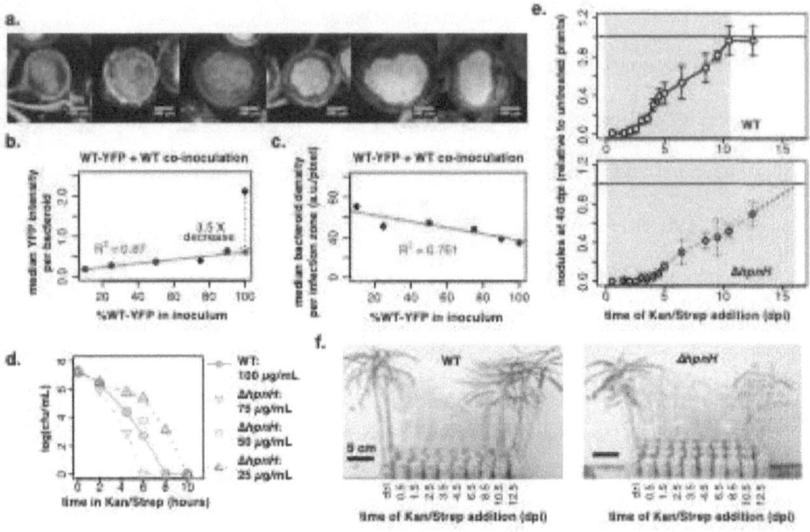

Figure 5. Extended hopanoid mutants are delayed at pre-intracellular stage(s) in symbiosis development. (a) Confocal sections of nodules from plants co-inoculated with wild type-YFP and Δ*hpnH*-mCherry harvested at 45-55 dpi. Sections were stained with Calcofluor (cyan) and are expressing YFP (yellow) and mCherry (magenta). (b) Scatter plot of median YFP intensity per pixel normalized by propidium iodide intensity per pixel (*e.g.* bacteroid density) within infection zones of nodules from plants co-inoculated with wild type-YFP and wild type, as a function of the percentage of wild type-YFP in the inoculum. (c) Scatter plot of median propidium iodide intensity per pixel (*e.g.* bacteroid density) within infection zones of nodules from plants co-inoculated with YFP-tagged wild type and untagged wild type, as a function of the percentage of WT-YFP in the inoculum. (d) Colony forming units/mL in wild-type and Δ*hpnH* cultures grown in BNM supplemented with varying concentrations of kanamycin and spectinomycin at various times post-inoculation. (e) Average nodules per plant at 40 dpi for plants inoculated with either wild type or Δ*hpnH* and treated with 50 µg/mL (Δ*hpnH*) or 100 µg/mL (wild type) kanamycin and streptomycin at various time points post-inoculation. Nodule counts are normalized to those observed in non-antibiotic treated plants. (f) Images of inoculated plants at 40 dpi after antibiotic treatment at various time points. Untreated plants are shown on the left, with increasing time of antibiotic addition. Error bars represent one standard deviation.

These effects of fluorophore overexpression made it difficult to interpret our WT-YFP and Δ*hpnH*-mCherry competition data, so we developed an alternative, antibiotics-based method to study the timing of early symbiotic initiation. First, we identified antibiotics that were effective against *B. diazoefficiens* but would minimally affect *A.*

afraspera growth. We tested three antibiotics (100 µg/ml streptomycin, 100 µg/ml kanamycin, and 20 µg/ml tetracycline) and treated non-inoculated plants with these antibiotics for two weeks, alone and in combination. After this treatment, we found that neither kanamycin nor streptomycin, nor the combination of the two, significantly affected plant appearance, shoot height, or root and shoot dry masses compared to untreated controls (**Fig. S19**). Plants treated with tetracycline were noticeably more yellow in color, indicating chlorosis, and the roots and plant medium became brown; these plants also had lower shoot and root dry masses than untreated plants (**Fig. S19**).

Because the Δ*hpnH* strain is more sensitive to antibiotics than wild type (Kulkarni et al. 2015), we tested various concentrations of the non-plant-perturbing antibiotics streptomycin and kanamycin to identify concentrations that would result in the same rates of cell death for both strains. We inoculated plant growth media with wild type or Δ*hpnH* to the same cell densities and under the same environmental conditions as in plant inoculation experiments. The wild-type culture was supplemented with 100 µg/ml streptomycin plus 100 µg/ml kanamycin, and Δ*hpnH* cultures were supplemented with decreasing concentrations of these antibiotics: 75, 50 and 25 µg/mL each. Samples of the cultures were collected, serially diluted and added to PSY plates to estimate colony-forming units (cfus) per mL over time. At 50 µg/mL kanamycin plus 50 µg/mL streptomycin, the rate of decrease in cfus/mL for Δ*hpnH* was equivalent to that of wild type treated with 100 µg/ml kanamycin plus streptomycin (**Fig. 5d**).

We inoculated 40 plants each with wild type or Δ*hpnH* and added streptomycin or kanamycin to 100 µg/mL each or 50 µg/mL each, respectively, at various points post-inoculation. After 40 days we counted the number of nodules per plant, and found that antibiotics were able to block nodule formation over a ~50% longer window in Δ*hpnH* compared to wild type (**Fig. 5e**). The decrease in nodules formed at different antibiotic treatment time points was also evident in the overall appearance of the plants (**Fig. 5f**). These results suggest that Δ*hpnH* requires more time on average to reach the intracellular stage of the symbiosis, at which point we presume that the bacteria are protected from the antibiotic by the host cells. These data would be consistent with Δ*hpnH* requiring more time to colonize the root surface, invade the root epidermis, and/or be internalized by host cells.

Extended hopanoids support surface attachment and motility in vitro

Because we found that expression of genetic tags in wild type and Δ*hpnH* perturbed their symbiosis with *A. afraspera,* and because we found that the hopanoid mutant viability is reduced by sonication, centrifugation, and mechanical or detergent-based tissue disruption techniques required to re-isolate bacteria from plants, we could not directly monitor the colonization of plant roots by these strains. Instead we used an *in vitro* approach to study the motility and adhesion of these strains on abiotic surfaces. Although these assays do not fully capture the environment of the root surface, defects in abiotic surface attachment can correlate with defects in host colonization (Nagy et al. 2015) and abiotic substrates can be sufficient to elicit some host response genes (Siryaporn et al. 2014). To determine whether Δ*hpnH* is less motile than wild type, we inoculated low-agar, PSY plates with Δ*hpnH* or wild type and measured the rate of zone of swimming over time. We observed that diameter of motility was reduced in Δ*hpnH* compared to wild type (**Fig. 6a-b**), consistent with a swimming motility defect; however, because we have previously shown that $\Delta hpnH$ grows more slowly in this medium than wild type (Kulkarni et al. 2015), we could not rule out the possibility that slower zone expansion simply reflects a longer doubling time.

To investigate the nature of the plate motility defect, we studied the motility of single *B. diazoefficiens* cells. We inoculated cells into a glass-bottom, sterile PSY flow cells with 100 µL of each strain and recorded the movement of cells near the glass surface at 5 ms time resolution (**Movies S1, S2**). Trajectories of individual motile cells, defined as cells having super-diffusive motion and a trajectory radius of gyration >2.5 µm, were calculated and analyzed in MATLAB (Lee et al. 2018). In agreement with results from motility plate assays, we observed significantly fewer ($p < 0.0001$) motile cells for in $\Delta hpnH$ cultures ($N = 65 \pm 29$) compared to wild-type cultures ($N = 368 \pm 60$) when cells were grown and assayed in PSY medium (**Fig. 6c-d; Table S1**). Among the motile cells in each population, average mean speeds did not differ significantly ($p>0.05$), with $<V>_{\Delta hpnH} = 24.83 \pm 7.0$ µm/sec and $<V>_{wt} = 22.75 \pm 6.7$ µm/sec. We repeated these assays in plant growth medium (BNM) supplemented with arabinose and ammonia (**Movies S3, S4**). Under this condition, we again observed a lower fraction of motile $\Delta hpnH$ cells than wild type ($N_{\Delta hpnH} = 54 \pm 59$ and $N_{wt} = 450 \pm 310$; $p < 0.01$) (Fig. 6e-f; Table S1). The mean speeds among motile cells grown and assayed in BNM were also similar between strains, with $<V>_{\Delta hpnH} = 25.04 \pm 6.6$ µm/sec and $<V>_{wt} = 22.99 \pm 6.4$ µm/sec, and did not differ significantly ($p>0.05$).

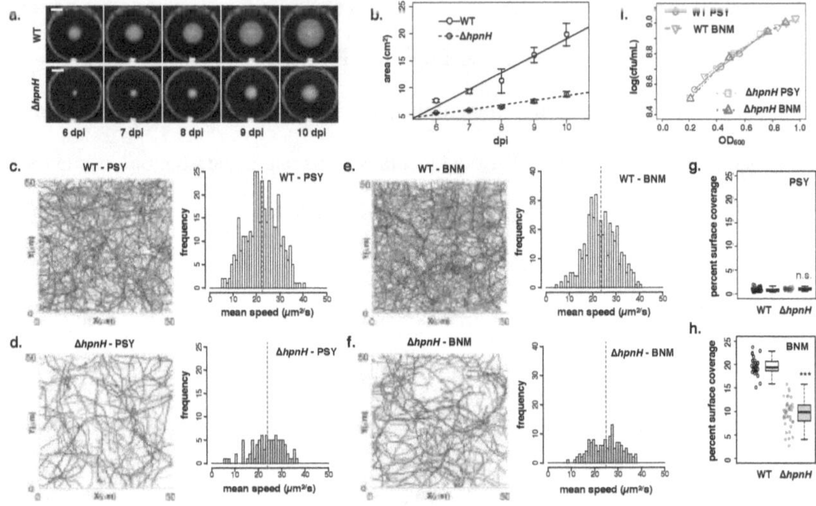

Figure 6. Extended hopanoid hutants are less motile than wild type and attach poorly to surfaces *in vitro*. (a) Sample time course of wild-type and Δ*hpnH* colony expansion on low-agar PSY plates (dpi = days post-inoculation). Scale bars represent 2 cm. (b) Average colony sizes of wild type and Δ*hpnH* over time. N=4 plates per strain; error bars indicate one standard deviation. (c) Mean speed distribution (N=359) and trajectories for motile wild-type cells observed over a 5 minute time course in PSY. (d) Mean speed distribution (N=91) and trajectories for motile Δ*hpnH* cells observed over a 5 minute time course in PSY. (e) Mean speed distribution (N=421) and trajectories for motile wild-type cells observed over a 5 minute time course in BNM. (f) Mean speed distribution (N=141) and trajectories for motile Δ*hpnH* cells observed over a 5 minute time course in BNM. Vertical dotted lines in the histograms shown in (c-f) indicate the distribution means. (g-h) Jitter and box plots of surface attachment (*e.g.* the percent of the field of view covered with cells) of WT and Δ*hpnH* after 2 hours of incubation on glass in (g) PSY or (h) BNM. N=40 fields of view per condition. Results of two-tailed t-tests between wild type and Δ*hpnH* are denoted as follows: n.s., p>0.05; ***, p<0.00001. (i) Colony forming units/mL in wild-type and Δ*hpnH* cultures grown to a range of OD_{600} in BNM supplemented with arabinose and ammonia or in PSY.

We tested the surface attachment capabilities of Δ*hpnH* and wild type by incubating dense bacterial cultures on glass coverslips and quantifying the fraction of the surface covered with stably adherent cells after two hours. In PSY medium, both strains adhered poorly, and there was no significant difference in their attachment efficiencies

(**Fig 6g; Fig. S20**). In BNM supplemented with arabinose, both strains adhered to glass better than in PSY, and Δ*hpnH* attachment levels were significantly lower than wild type (**Fig. 6h; Fig. S20**). Because a decrease in stably adherent cells and in motile cells within Δ*hpnH* cultures could reflect a lower number of viable cells in this strain, we also measured colony forming units per mL in wild-type and Δ*hpnH* cultures grown to varying cell densities (OD_{600} 0.2-1.0) in either PSY or BNM supplemented with arabinose and ammonia. We did not find differences in the cfus/mL in each strain for any of the cell densities and medium conditions tested (**Fig. 6h**), demonstrating that reduced *in vitro* adhesion and motility among Δ*hpnH* cells cannot be attributed to higher levels of cell death. The decreased adhesion and reduced motile cell population of Δ*hpnH* suggest that stable root colonization by this strain may be less efficient, although we cannot account for possible differences in adhesion mechanisms used during attachment to glass versus attachment to the plant root surface.

Discussion

Hopanoids are well-established mediators of bacterial survival under stress, and previously we showed that the capacity for hopanoid production is enriched in plant-associated environments (Ricci et al. 2014) and required for optimal *Bradyrhizobia-Aeschynomene* spp. symbioses (Silipo et al. 2014; Kulkarni et al. 2015). Here we performed a detailed, quantitative evaluation of the extended hopanoid phenotypes in the *Bradyrhizobium diazoefficiens-Aeschynomene afraspera* symbiosis. We determined that extended hopanoid mutants fix nitrogen at similar rates as wild type on a per-bacteroid level, demonstrating that in this host extended hopanoids are not required to protect nitrogenase from oxygen, as often has been speculated (reviewed in Belin et al. 2018). Instead, we found that the extended hopanoid mutants' lower *in planta* productivity can be fully attributed to changes in the kinetics of nodule development. By tracking the development of individual root nodules, we observed later nodule emergence times in Δ*hpnH*-inoculated plants. *In vitro*, Δ*hpnH* cells adhered poorly to glass and were less motile than wild type, and it is possible that Δ*hpnH* are similarly deficient in motility and adhesion in the context of plant association, leading to slower attachment to plant root surfaces (**Fig. 7a-b**). While slower root attachment and/or reduced motility toward and within lateral root-associated cracks could explain the later emergence times of Δ*hpnH* nodules, more experiments will be needed to determine whether our *in vitro* results are relevant to the native *Bradyrhizobium-Aeschynomene* association.

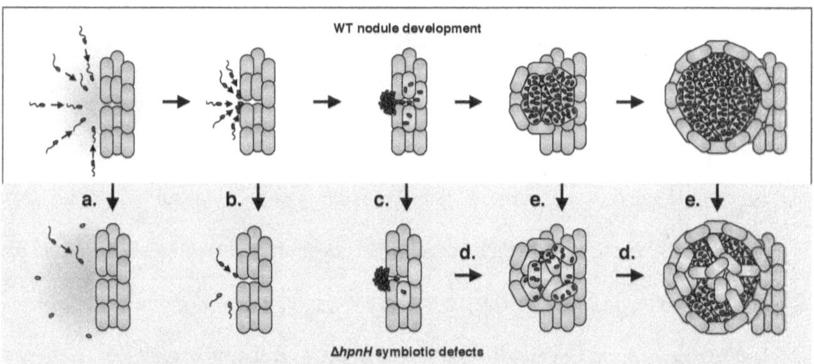

Figure 7. Consequences of extended hopanoid loss in *A. afraspera* nodule development. Schematic representation of *A. afraspera* wild-type root nodule development (top row; white background) and defects in development associated with extended hopanoid loss (bottom row; grey background). Early in development, fewer Δ*hpnH* cells are motile (a) and competent to attach to root surfaces (b), leading to a delay in establishment of stable root colonies. At later stages, slow growth of Δ*hpnH* into the root interior, or poor uptake by and division within host cells (c) may generate "patchy" or under-populated infection zone that is propagated as the nodule grows (d). Alternately, fully infected Δ*hpnH* nodules may lose symbionts to symbiont cell death (e) via poor bacteroid survival or plant-directed symbiosome degradation.

Through our developmental tracking, we also found that one third of the Δ*hpnH* nodules also grew significantly slower than wild type and were smaller at maturity. Many of these small nodules contained low symbiont densities; a subset of larger Δ*hpnH* nodules also had lower symbiont loads, due to infection zone fragmentation. The origin of this under-infection is unclear. It is possible that bacteria are inefficiently internalized or retained, and this phenotype is simply propagated as nodules develop (**Fig. 7c-d**). Alternatively, low symbiont densities may reflect symbiont degradation in a previously fully infected nodule (**Fig. 7e**), perhaps correlating with elicitation of a plant defense response.

These observations challenge two conclusions from our previous work, requiring a refinement of our interpretation of the roles of extended hopanoids in the plant context (Kulkarni et al. 2015). First, we reported that there was no symbiotic defect of the Δ*hpnH* strain in soybean, based on the observation that nitrogen fixation per mg nodule dry weight was similar to wild type. Given that this study revealed that a reduction in nodule dry weight explains the Δ*hpnH* defect in *A. afraspera*, it is possible that this strain is also defective in soybean, but this defect was obscured by differences in normalization between the soybean and *A. afraspera* datasets. Second, the majority of Δ*hpnH* nodules in *A. afraspera* had wild-type-like growth kinetics and morphologies, with a few "mega" nodules displaying unusually fast growth. This finding appears inconsistent with an inability to survive *A. afraspera* NCR peptides, unless NCR peptide expression levels in *A. afraspera* are extremely variable from nodule to nodule, or if the mechanisms that compensate for extended hopanoid loss are inconsistent.

What alternate mechanisms might underpin these extended hopanoid mutant phenotypes? Perhaps they are simply consequences of less rigid *B. diazoefficiens* membranes. The fraction of motile cells in *E. coli* populations has been suggested to be sensitive to changes to the mechanical properties of the outer membrane (Gupta et al. 2006), and membrane-based mechanotransduction is required by diverse bacteria to stimulate extracellular matrix production and cement their attachment to surfaces (Petrova and Sauer 2012; Persat 2017). *B. diazoefficiens* mutants with weakened cell walls also have been shown to be deficient in symbiosis with *A. afraspera* through an NCR peptide-independent mechanism (Barriere et al. 2017), which may be elicited by *ΔhpnH*.

Extended hopanoid loss may also have secondary effects on *Bradyrhizobium-Aeschynomene* signaling. In the *Frankia*-actinorhizal symbiosis, bacterial extended hopanoids can contain the auxinomimetic compound phenyl-acetic acid (PAA) (Hammad et al. 2003), and though the effects of hopanoid loss on the bacterial metabolome have not been examined, changes in hopanoid production may impact the synthesis and/or secretion of symbiotically active compounds. Future work will be required to determine whether changes in signaling or membrane mechanics dominate the hopanoid mutant phenotypes, and at which developmental stages.

Regardless of the underlying mechanism, it is curious that the absence of extended hopanoids is not a death knell for the *B. diazoefficiens-A. afraspera* symbiosis at any stage. In our *in vitro* studies, mean speeds among motile *ΔhpnH* cells were indistinguishable from wild type, and though we cannot rule out more subtle defects in the direction of movement or chemotaxis, this suggests that motility systems of *ΔhpnH* cells function properly once induced. Similarly *in planta*, Δ*hpnH* nodules developing at wild-type rates and reaching average wild-type volumes did occur–and, in the case of "mega" nodules, some exceeded their wild-type counterparts.

Why do Δ*hpnH* populations form two distinct populations (wild-type-like or defective) rather than falling on a continuous distribution of behavior? Bimodality can reflect switch-like, or threshold-based, regulation, and perhaps in the Δ*hpnH* strain, a fraction of cells cannot support levels of signaling above the threshold required for proper function. Nodules may also differ in the extent to which extended hopanoid loss is compensated. In *Methylobacterium extorquens* and *Rhodopseudomonas palustris* (Bradley et al. 2017; Neubauer et al. 2015), hopanoid loss results in upregulation of other membrane-rigidifying lipids including carotenoids and cardiolipins, and in other plant-microbe systems, lipid exchange between hosts and microbes has been observed (Keymer 2018), suggesting that Δ*hpnH* nodule phenotypes may relate to the local availability of structurally or functionally similar metabolites. Because of these diverse possible explanations for Δ*hpnH* heterogeneity, a detailed comparison of wild-type-like and defective nodules, including the distributions of lipids and other metabolites, bacteroid morphology and penetrance, and gene expression variability, will be required to determine why some Δ*hpnH* nodules succeed and others do not.

Beyond hopanoids, our results provide insight into the developmental control of nodule formation by *A. afraspera* hosts. We find that nodulation occurs in bursts separated by fixed 18-day intervals, and that the timing of these bursts is unrelated to net fixed nitrogen production across the root, more likely reflecting the inherent dynamics of the underlying signaling networks. The growth period of individual nodules is similarly deterministic, suggesting that *A. afraspera* hosts do not respond to ineffective symbionts by prematurely aborting nodule development. Rather, we find that *A. afraspera* nodules can be primarily distinguished by their growth rates, *e.g.* the frequencies of infected host cell division. This finding suggests that in *A. afraspera* host-cell mitosis and symbiont performance may be coupled, enabling future studies on the molecular signals through which this coupling occurs.

Finally, our results underscore the importance of identifying the most informative, least perturbing tools for interrogating legume-microbe symbiosis. Employing quantitative, time-resolved, single-nodule and single-cell approaches rather than bulk measurements were essential for uncovering the diverse phenotypes of the *B. diazoefficiens* extended hopanoid mutants and yielded unexpected information on regulation of nodule development by *A. afraspera*. We have also shown the limitations of introducing overexpressed genetic tags into bacteria. While use of these tags has undoubtedly enhanced our understanding of legume-microbe symbiosis (Ledermann et al. 2018), they may not fully capture the behavior of native organisms. Additionally, our work is one of many to emphasize the importance of appropriate culture models for mimicking the host environment, as the Δ*hpnH* surface attachment defect was observed in plant growth medium but not in a standard richer medium. A more detailed analysis of the host environment, including the full milieu of root exudates (Sugiyama and Yazaki 2012), available carbon sources (Pini et al. 2017), and trace metals specific to each legume, will improve *in vitro* models of legume-bacteria interactions and may allow selection of strains with improved performance in agriculture.

Methods

B. diazoefficiens *culture and strain generation*

   B. diazoefficiens hopanoid biosynthesis mutants were generated previously

(Kulkarni et al. 2015). For construction of YFP- and mCherry-expressing strains,

fluorophore expression vectors pRJPaph-YFP and pRJPaph-mCherry (Ledermann et al.

2015) were provided as a gift from Prof. Dr. Hans-Martin Fischer (ETH Zurich). For

complementation of Δ*hpnH* with the endogenous *hpnH* gene on the strong *Paph*

promoter of pRJPaph-mCherry, mCherry was replaced with *Paph-hpnH*. These vectors

were introduced into *B. diazoefficiens* by conjugation with the β2155 DAP auxotroph

strain of *E.coli*, using the following protocol: *B. diazoefficiens* wild type and Δ*hpnH* were

grown in 5 mL PSY medium (Regensburger and Hennecke, 1983) at 30°C and 250 rpm

to an OD_{600} of ~1.0 (wild type) or of 0.5-0.8 (Δ*hpnH*). β2155 strains carrying pRJPaph

vectors were grown to an OD_{600} of 0.5-0.8 in 5 mL LB supplemented with 10 μg/mL

tetracycline and 300 μm DAP at 37°C and 250 rpm. Both *B. diazoefficiens* and β2155

donor cultures were pelleted at 3250 x g for 30 minutes, washed three times in 0.9%

sterile saline, and resuspended in 0.9% sterile saline to a final OD_{600} of 1.0. *B.*

diazoefficiens strains and β2155 donor cells were combined at a 4:1 ratio, respectively,

and mixed by repeated pipetting. Aliquots (50 μl) of these 4:1 mixtures were dropped to

PSY plates supplemented with 300 μm DAP, dried in a biosafety cabinet, and incubated

for 48 hours at 30°C. Conjugation pastes were removed from plates and resuspended in 5

mL sterile saline, pelleted at 3250 x g for 30 minutes and washed twice, in order to

remove residual DAP. Washed cells were pelleted a final time and resuspended to 200 μl in 0.9% sterile saline and plated onto PSY plates supplemented with 20 µg/mL (wild type) or 10 µg/mL (Δ*hpnH*) tetracycline. Colonies appeared after 7-10 days (wild type) or 10-14 days (Δ*hpnH*) and were streaked onto fresh PSY/tetracycline plates, and sequenced to verify insertion of the pRJPaph vectors into the *scoI* locus.

A. afraspera *cultivation and inoculation with* B. diazoefficiens

A. afraspera seeds were obtained as a gift from the laboratory of Dr. Eric Giraud (LSTM/Cirad, Montpelier, France). Seeds were sterilized and scarified by incubation in 95% sulfuric acid at RT for 45 minutes, followed by 5 washes in sterile-filtered nanopure water and a second incubation in 95% ethanol for 5 minutes at RT. After ethanol treatment seeds were washed 5X and incubated overnight in sterile-filtered nanopure water. Seeds were transferred to freshly poured water/agar plates using sterile, single-use forceps in a biosafety cabinet, and germinated for 24-72 hours in the dark at 28-32°C.

Seedlings were placed in clear glass test tubes containing 100 mL of sterile, nitrogen-free Buffered Nodulation Medium (BNM) (Ehrhardt et al. 1992) and grown for 7-10 days in plant growth chambers (Percival) under the following settings: 28°C, 80-90% humidity, and 16 hour photoperiod under photosynthetic light bulbs (General Electric) emitting ~4000 lumens/ft^2. In parallel, *B. diazoefficiens* strains were grown in 5-10 mL PSY liquid culture at 30°C and 250 rpm to stationary phase (OD_{600} > 1.4). Stationary phase cultures were diluted into PSY one day prior to plant inoculation to reach an OD_{600} of ~0.8 at the time of inoculation. OD_{600} ~ 0.8 cultures were pelleted at 3250 x g for 30 minutes at RT, washed once in PSY, and resuspended in PSY to a final OD_{600} of 1.0. Resuspended *B. diazoefficiens* cultures were directly inoculated into the plant medium in a sterile biosafety cabinet; 1 mL of OD_{600}=1.0 culture was added per plant. Inoculated plants were returned to growth chambers and maintained for the times indicated for each experiment. For longer experiments (lasting longer than ~30 days post-inoculation), plant growth tubes were refilled with sterile-filered nanopure water as needed. To minimize cross-contamination, inoculated plants and non-inoculated plants were cultivated in separate growth chambers, and growth chambers were sterilized with 70% ethanol followed by UV irradiation for at least 24 hours between experiments.

Acetylene reduction experiments

Individual plants were transferred to clear glass 150 mL Balch-type anaerobic culture bottles containing 15 mL BNM medium and sealed under a gas-tight septum. After sealing, 15 mL of headspace gas (10% of the culture bottle volume) was removed and replaced with 15 mL of acetylene gas (Airgas). Plants in culture bottles were incubated in the light at 28°C in growth chambers for 3-6 hours. A 100 μl sample of the headspace gas was removed using a gas-tight syringe (Hamilton), and this sample was injected and analyzed for ethylene signal intensities using a Hewlett Packard 5890 Series II GC with Hewlett Packard 5972 Mass Spectrometer with a 30mx0.320mm GasPro Column (Agilent Technologies) and a 2 mm ID splitless liner (Restek Corporation). Following acetylene reduction measurements, plants were removed from jars and plant shoot heights and number of nodules per plant were recorded. When nodule dry mass measurements were performed, nodules were harvested with a razor blade, transferred into pre-weighed Eppendorf tubes, dried at 50°C for a minimum of 48 hours, and weighed again.

Live:Dead staining and imaging of nodule cross-sections

Nodules were hand-sectioned with razor blades or on a Leica VT1200 vibratome and immediately transferred into a fresh solution of 5 µM SYTO9 (diluted 1:100 from a 500 uM stock in DMSO at -20°C; Thermo Fisher) and 0.02 mg/mL (30 µM) propidium iodide (diluted 1:50 from a 1 mg/mL stock stored in water at 4°C; Thermo Fisher) in PBS. Nodule sections were incubated in this SYTO9/propidium iodide solution at room temperature for 30 minutes in the dark with gentle shaking, washed 5X in PBS, and fixed in 4% paraformaldehyde (Electron Microscopy Sciences) in PBS overnight in the dark at 4°C. Fixed sections were washed 5X in PBS and transferred to a freshly prepared solution of 0.1 mg/mL Calcofluor White (Fluorescence Brightener 28; Sigma) in PBS. The sections were incubated in the Calcofluor solution in the dark for 1 hour at RT with gentle shaking and washed 5X in PBS to remove excess dye.

Prior to imaging, sections were transferred to 30 mm imaging dishes with 20 mm, #0 coverglass bottoms (MatTek) and overlaid with sterile 50% glycerol. Nodule images were collected on either a Leica TCS SPE laser-scanning confocal (model DMI4000B-CS) using a 10X/0.3 NA APO ACS objective (for low-resolution images) or a Zeiss LSM 880 laser-scanning confocal equipped with a Fast Airyscan super-resolution module using a 63X/1.25 NA Plan-Neofluar objective (for high-resolution images). Fluorophore excitation was performed at the following settings for each dye: Calcofluor, 405 nm excitation/410-500 nm emission; SYTO9, 488 nm excitation/510-570 nm emission; PI, 532 nm excitation/600-650 nm emission. These images were processed to enhance brightness and contrast in FIJI (Schindelin et al. 2012; Schneider et al. 2012).

Nodule diameter and volume measurements

Inoculated *A. afraspera* root nodules were imaging using a high-definition Keyence VHX-600 digital microscope at 20X magnification. For end-point root nodule volume measurements at 40 days post-inoculation, plants were removed from the growth chamber and imaged at RT on paper towels, and discarded. Nodule diameters were measured using the line tool in FIJI and recorded using a custom FIJI macro. For tracking nodule volumes over time, plants were serially removed from their growth chambers and transferred to a plastic dish containing 150 mL of sterile BNM pre-warmed to 28°C. Images of sections of the plant root were collected serially from the hypocotyl to the root tip. Following collection of images, plants were immediately returned to their original growth tubes in the growth chamber. Plastic dishes were sterilized for 10 minutes in 10% bleach, washed three times in sterile-filtered nanopure water, sprayed with 70% ethanol/water, and air-dried before each new plant was imaged. A fresh aliquot of sterile, pre-warmed BNM also was used for each plant. After the time course was completed, images of entire plant root systems were reconstructed by eye for each plant at each time point. For nodules appearing in at least five time points, nodule diameters were measured as described for the end-point measurements and were converted to approximate volumes in R using the equation $V = 4/3\pi r^3$.

Nodule growth curve fitting and analysis

All analyses of nodule growth, and corresponding plots, were generated in R. For nodule growth curve fitting, three model equations were used to identify the best fit, as follows:

(1) exponential function:

$$V = ae^{-bt} + c$$

(2) quadratic function:

$$V = at^2 + bt + c$$

(3) generalized logistic function (expressed as a Richard's function with a time shift):

$$V = \frac{a}{(1 + e^{-b(t-c)})^{\left(\frac{1}{d}\right)}}$$

Calculation of the optimal parameter values for each equation (e.g. the values of **a, b, c,** and **d**) and the standard error for each curve compared to the raw data were performed using the built-in function *nlm*() in R. In some cases, *nlm()* could not produce a best-fit model without specifying initial values for the function parameters. For exponential models, an equation of best fit could be successfully determined without specification of initial values for parameters **a, b,** and **c.** For quadratic models, initial parameter values were required and were set to **a**=0, **b**=10 and **c**=0 for each nodule plot, after identifying these initial parameter values as broadly optimal based on an initial parameter sweep of -50 to 50 for each plot. For sigmoidal models, no broadly optimal initial values could be identified, so a parameter sweep was performed for each plot with the initial value of **a** set to the maximum observed nodule volume (as **a** describes the upper asymptote of the sigmoidal curve), **b** ranging from 0.1 to 1, **c** ranging from 0 to 10, and **d** ranging from 0.01 to 1.0. In the sigmoidal plots, an initial point of (0,0) was added to the nodule volume time series to improve fitting.

Because the sigmoidal model provided the best fits, extrapolation of nodule growth characteristics was performed on sigmoidal models only. The maximum nodule volume, V_{max}, is defined as the upper asymptote of the sigmoidal growth curve, *e.g.* **a**. The nodule initiation time, t_{min}, was defined in three separate ways: the times at which the nodule volume is equal to 0.05, 0.1, or 0.2 mm^3 (*e.g.* through solving 0.05, 0.1, or 0.2 = $\mathbf{a}/((1+e^{(-\mathbf{b}(t-\mathbf{c}))})^{(1/\mathbf{d})})$ for t). The maximum nodule growth rate, dV/dt, was defined as the average rate of growth (*e.g.* slope) between the time at which the volume is 10% of V_{max} and the time at which the volume is 90% of V_{max}. The time at which each nodule reaches its maximum size, t_{max}, was approximated as the time at which the volume is 90% of V_{max}, since the "true" maximum volume is asymptotic to the growth curve and is therefore never fully reached in the model.

Competition assays

mCherry-tagged Δ*hpnH* and YFP-tagged wild-type *B. diazoefficiens* were grown to stationary phase (OD_{600} > 1.4) in 10 mL PSY cultures supplemented with 20 µg/mL (wild type) or 10 µg/mL (Δ*hpnH*) tetracycline; untagged strains were grown in PSY. On the day prior to inoculation, all strains were diluted into 50-150 mL tetracycline-free PSY to reach an OD_{600} of ~0.8 at the time of inoculation. *A. afraspera* plants were cultivated pre-inoculation in test tubes as described above, with the addition of covering the growth tubes in foil to minimize the production of chlorophyll in the plant roots, which spectrally overlaps with mCherry. At the time of inoculation, all cultures were pelleted at 3250 x g for 30 minutes at RT, washed three times, and resuspended in PSY to a final OD_{600} of 1.0. A 10 mL culture of each strain ratio for inoculation was generated a sterile 15mL Falcon tube; for example, for a 50:50 mixture of mCherry-tagged $\Delta hpnH$ and YFP-tagged wild type, 5 mL of each strain was combined. These cultures were mixed thoroughly by gentle pipetting, and 1 mL of the mixtures was added to directly to the plant medium for 7-8 plants per strain mixture.

After 45-60 days, plants were harvested. First, plant heights and the number of nodules per plant were recorded. The roots were cut from the stem and images of all nodules for each plant were collected on a high-definition Keyence VHX-600 digital microscope at 20X magnification. These nodules were cross-sectioned and immediately transferred to Eppendorfs containing 4% paraformaldehyde (Electron Microscopy Sciences) in PBS. Fresh sections were fixed overnight in the dark at 4°C, washed 5X in PBS, and stored in PBS supplemented with 0.1% azide in the dark at 4°C until imaging.

Fixed sections were stained in Calcofluor (all strain combinations), SYTO9 (WT-YFP and WT co-inoculation only) or propidium iodide (mCherry-$\Delta hpnH$ and $\Delta hpnH$ co-inoculation only) as described for Live:Dead staining. Imaging was performed as described for Live:Dead staining using a 5X objective. Given the high autofluorescence of these nodules and low mCherry and YFP signal intensities, the following excitation/emission settings were used: Calcofluor, 405 nm excitation/410-460 nm emission; YFP/SYTO9, 488 nm excitation/500-550 nm emisasion; mCherry, 532 nm excitation/600-650 nm emission.

Quantification of nodule statistics (including nodule and infection zone areas, signal intensity of YFP, mCherry, SYTO9 and propidium iodide) was performed on raw images using a custom FIJI macro. Briefly, nodule images were opened at random, infection zones (IZs) and whole nodules were circled by hand and saved as discrete regions of interest (ROIs), and the area and intensity in each channel were measured automatically for all ROIs. These measurements were exported as a text table and various parameters from these measurements were calculated using custom Python scripts, as indicated in the Results. Plots of all parameters and statistical comparisons were generated using custom R scripts. All custom scripts and data used for this analysis are available upon request.

Antibiotic treatment of inoculated plants

A. afraspera plants were cultivated as described above, and the following antibiotics were added to non-inoculated plants 7 days after rooting in 100 mL BNM growth tubes: kanamycin to 100 µg/mL, streptomycin to 100 µg/mL, tetracycline to 20 µg/mL, kanamycin plus tetracycline, kanamycin plus streptomycin, streptomycin plus tetracycline. Plants were grown in antibiotics under normal plant growth conditions for 14 days, after which plants were visually inspected. Plant heights were also recorded, and the root and shoot systems were separated with a razor blade, transferred into pre-weighed 15 mL Falcon tubes, dried at 50°C for a minimum of 48 hours, and weighed again.

Antibiotic treatments of $\Delta hpnH$ and wild-type *B. diazoefficiens* were performed by growing antibiotic 5 mL PSY cultures of each strain to stationary phase (OD_{600} >1.4) and diluting strains in fresh PSY to reach an OD_{600} of ~0.8 at the time of antibiotic treatment – *e.g.* as they would be grown prior to plant inoculation. Cultures were pelleted at 3250 x g for 30 minutes at RT, washed three times, and resuspended in PSY to a final OD_{600} of 1.0. Four 100 µl aliquots of these culture were diluted 1:00 into separate 10 mL BNM cultures in clear glass tubes in plant growth chambers. Kanamycin (at 25, 50, 75, and 100 µg/mL) and streptomycin (at 25, 50, 75, and 100 µg/mL) were added directly to the BNM cultures, and 100 µl samples were taken immediately prior to antibiotic treatment and at 2, 4, 6, 8, and 10 hours post-antibiotic addition. These 100 µl samples were immediately diluted 1:10 in 900 µl and mixed vigorously by repeated pipetting. Vortexing was avoided as we found that this method reduces $\Delta hpnH$ viability. Ten serial 1:10 dilutions were performed, and three 10 µl samples of each dilution for each strain were spotted and dripped across PSY plates. After 7 days (wild type) or 10 days ($\Delta hpnH$), colonies were counted manually and recorded for each dilution exhibiting discrete colonies. Log plots of colony counts over time were generated in R.

Plants were inoculated with $\Delta hpnH$ and wild-type *B. diazoefficiens* as described above, and kanamycin and streptomycin were added to $\Delta hpnH$-inoculated plants to 50 µg/mL each, and to wild-type-inoculated plants to 100 µg/mL at 12 hours and 36 hours and at 2, 2.5, 3, 3.5, 4, 4.5, 5, 6.5, 8.5, 9.5, 10.5, and 12.5 days post-inoculation. Four plants were treated per time point per strain, with an additional four plants each as an untreated control. At 40 dpi, the number of nodules per plant was recorded.

Bulk motility assays

Swimming motility assays were performed as previously described, with some modifications (1). WT and $\Delta hpnH$ were grown to turbidity in 5 mL of PSY at 30°C and 250 rpm, diluted to an OD_{600} of 0.02 in 5 mL of fresh PSY, and grown to exponential phase (OD_{600} = 0.3-0.5). Exponential cultures were diluted to an OD_{600} of 0.06 in fresh PSY and 2 μL of the adjusted cultures into the center of swimming plate containing 0.3% agar/PSY. After inoculation, the plates were wrapped with parafilm to prevent dehydration and incubated in a humidity-controlled environmental chamber (Percival) at 30°C for 10 days total, with daily scans after 5 days. The resulting images were analyzed in FIJI to measure the area of the swimming colony.

Surface attachment assays

$\Delta hpnH$ and wild-type *B. diazoefficiens* were grown in 5 mL PSY cultures to stationary phase (OD_{600} >1.4) diluted in fresh PSY to reach an OD_{600} of ~0.8 at the time of surface attachment assays. Cultures were pelleted at 3250 x g for 30 minutes at RT, washed twice in the indicated attachment medium, and resuspended in attachment medium to an OD_{600} of 1.0. These cultures were mixed thoroughly by repeated pipetting, and 2 mL samples were added to sterile imaging dishes (30 mm dishes with 20 mm, #1.5 coverglass bottoms; MatTek). Cultures were incubated on imaging dishes *without shaking* at 30°C for two hours. To remove non-adhered cells, imaging dishes were immersed in 50 mL of attachment media in a 100 mL glass beaker on an orbital shaker and shaken gently at RT for 5 minutes; direct application of washing medium to the coverglass surface was avoided, as we found that this creates a shear force sufficient to wash away adhered cells. Imaging dishes were gently lifted out of the washing medium and imaged with a 100X objective on a Lumascope 720 fluorescence microscope (Etaluma). Forty fields of view were recorded for each strain and media combination. These images were processed in FIJI using the Enhanced Local Contrast (CLAHE) plugin (Heckbert and Karel 1994) and converted into a binary image to determine the area of the imaging window covered with adhered cells. Calculation of the fraction of the surface was performed in Excel and statistical analyses were conducted in R. Areas of the surface containing groups of cells larger than 10 μm^2 in area were ignored in the calculations, as these likely do not represent true attachment events rather than sedimentation of larger cell clumps. BNM used for attachment assays was prepared as described above, with the addition of 1.0 g/mL arabinose. Because BNM contains salt crystals that can sediment onto coverglass and occlude or obscure adhered cells, this medium was passed through a 2 μm filter (Millipore) prior to the attachment experiments.

Single-cell motility assays and analysis

B. diazoefficiens wild type and ΔhpnH were grown in 12.5 ml PSY medium at 30°C and 200 rpm to an $OD_{600} = 0.6$-0.8 from an AG medium plate culture. A 1:10 dilution of cell culture was subcultured in PSY medium to a final volume of 12.5 ml and regrown to an OD_{600} of ~0.6. Two aliquots of 750 μL were sampled from the regrowth culture and pelleted at 3500 x g for 20 min (wild type) or for 30 min (ΔhpnH) at RT. The supernatant was removed, and one pellet was resuspended in 500 μL PSY and the other in 500 μL BNM medium. Because BNM contains salt crystals that can sediment onto coverglass and occlude or obscure adhered cells, this medium was passed through a 2 μm

filter (Millipore) prior to usage for these experiments. The two medium conditions were incubated for 2.5 hrs (wild type) or for 3.5 hrs (ΔhpnH) at 30°C; given the difference in growth time ΔhpnH incubated for longer. Right before imaging, each culture was diluted at a 1:10 ratio with its respective medium. The bacteria were injected into a sterile flow cell (ibidi sticky-Slide VI0.4 with a glass coverslip). The flow cell was attached to a heating stage set to 30°C.

The imaging protocol involved high-speed bright-field imaging for 5 min at a single XYZ location per experimental repeat. High speed bright-field recordings used a Phantom V12.1 high speed camera (Vision Research); images were taken with a 5 ms exposure at 200 fps and a resolution of 512×512 pixels (0.1 μm/pixel). This protocol was performed on an Olympus IX83 microscope equipped with a $100\times$ oil objective, a $2\times$ multiplier lens, and a Zero Drift Correction autofocus system. The recorded movies were extracted into single frames from the .cine files using PCC 2.8 (Phantom Software). Image processing and cell tracking algorithms are adapted from previous work (Lee et al. 2018) and written in MATLAB R2015a (Mathworks).

We identified cells swimming near the surface as cells with a trajectory radius of gyration greater than 2.5 μm and a mean-squared displacement (MSD) slope greater than 1.5. Setting a minimum radius of gyration selects for cells with a minimum net translation on the across the surface, while a minimum MSD slope threshold ensured the cells are moving super-diffusively (MSD slope $\cong$ 1, diffusive motion; MSD slope $\cong$ 2, super-diffusive motion). For each tracked cell, the mean-speed, **v**, was calculated by averaging a moving window, **w**, of the displacement over the cell's full trajectory, using the following equation:

$$< v > = \; Avg\left(\sum_{t=1}^{N-w} \frac{\sqrt{(x_{t+w} - x_t)^2 + (y_{t+w} - y_t)^2}}{w} * f * p\right)$$

where **N** is the total number of points in the trajectory, **f** is the acquisition frame rate, and **p** is the pixel resolution. Here we set a window size, w= 40 frames. All analysis and visualizations from these experiments where done using MATLAB R2015a (Mathworks).

Data Availability

Supplementary tables and movies referred to but not included in this document due to their size and/or format are available through CaltechDATA and are linked to the record of this thesis in CaltechTHESIS.

References

Abeysekera, R. M., Newcomb, W., Silvester, W. B., and Torrey, J. G. 1990. A freeze-fracture electron microscopic study of *Frankia* in root nodules of *Alnus incana* grown at three oxygen tensions. Can. J. Microbiol. 36:97–108
Article CARUNA, R. *PHARMACOGNOSTIC STUDIES OF AESCHYNOMENE INDICA L.*
Barriere, Q., Guefrachi, I., Gully, D., Lamouche, F., Pierre, O., Fardoux, J., Chaintreuil, C., Alunni, B., Timchenko, T., Giraud, E., and Mergaert, P. 2017. Integrated roles of BclA and DD-carboxypeptidase 1 in Bradyrhizobium differentiation within NCR-producing and NCR-lacking root nodules. Sci. Rep. 7:9063
Belin, B. J., Busset, N., Giraud, E., Molinaro, A., Silipo, A., and Newman, D. K. 2018. Hopanoid lipids: from membranes to plant-bacteria interactions. Nat. Rev. Microbiol. 16:304–315
Bonaldi, K., Gargani, D., Prin, Y., Fardoux, J., Gully, D., Nouwen, N., Goormachtig, S., and Giraud, E. 2011. Nodulation of Aeschynomene afraspera and A. indica by photosynthetic Bradyrhizobium Sp. strain ORS285: the nod-dependent versus the nod-independent symbiotic interaction. Mol. Plant. Microbe. Interact. 24:1359–1371
Bradley, A. S., Swanson, P. K., Muller, E. E. L., Bringel, F., Caroll, S. M., Pearson, A., Vuilleumier, S., and Marx, C. J. 2017. Hopanoid-free Methylobacterium extorquens DM4 overproduces carotenoids and has widespread growth impairment. PLoS One. 12:e0173323
Caamal-Fuentes, E., Peraza-Sánchez, S., Torres-Tapia, L., and Moo-Puc, R. 2015. Isolation and Identification of Cytotoxic Compounds from Aeschynomene fascicularis, a Mayan Medicinal Plant. Molecules. 20:13563–13574
Chifundera, K. 2001. Contribution to the inventory of medicinal plants from the Bushi area, South Kivu Province, Democratic Republic of Congo. Fitoterapia. 72:351–368
Czernic, P., Gully, D., Cartieaux, F., Moulin, L., Guefrachi, I., Patrel, D., Pierre, O., Fardoux, J., Chaintreuil, C., Nguyen, P., Gressent, F., Da Silva, C., Poulain, J., Wincker, P., Rofidal, V., Hem, S., Barrière, Q., Arrighi, J.-F., Mergaert, P., and

Giraud, E. 2015. Convergent Evolution of Endosymbiont Differentiation in Dalbergioid and Inverted Repeat-Lacking Clade Legumes Mediated by Nodule-Specific Cysteine-Rich Peptides. Plant Physiol. 169:1254–1265

Ehrhardt, D. W., Atkinson, E. M., and Long, S. R. 1992. Depolarization of alfalfa root hair membrane potential by Rhizobium meliloti Nod factors. Science. 256:998–1000

Gupta, R., Sharma, M., and Mittal, A. 2006. Effects of membrane tension on nanopropeller driven bacterial motion. J Nanosci Nanotechnol. 6:3854–3862

Hammad, Y., Nalin, R., Marechal, J., Fiasson, K., Pepin, R., Berry, A. M., Normand, P., and Domenach, A.-M. 2003. A possible role for phenyl acetic acid (PAA) on Alnus glutinosa nodulation by Frankia. Plant Soil. 254:193–205

Heckbert, P. S., and Karel. 1994. *Graphics gems IV*. AP Professional.

Keymer, A. 2018. Cross-kingdom lipid transfer in arbuscular mycorrhiza symbiosis and beyond. Curr. Opin. Plant Biol. 44:137–144

Komaniecka, I., Choma, A., Mazur, A., Duda, K. A., Lindner, B., Schwudke, D., and Holst, O. 2014. Occurrence of an unusual hopanoid-containing lipid A among lipopolysaccharides from Bradyrhizobium species. J. Biol. Chem. 289:35644–35655

Kulkarni, G., Busset, N., Molinaro, A., Gargani, D., Chaintreuil, C., Silipo, A., Giraud, E., and Newman, D. K. 2015. Specific hopanoid classes differentially affect free-living and symbiotic states of Bradyrhizobium diazoefficiens. MBio. 6:e01251-15

Ledermann, R., Bartsch, I., Muller, B., Wulser, J., and Fischer, H.-M. 2018. A Functional General Stress Response of Bradyrhizobium diazoefficiens Is Required for Early Stages of Host Plant Infection. Mol. Plant. Microbe. Interact. 31:537–547

Ledermann, R., Bartsch, I., Remus-Emsermann, M. N., Vorholt, J. A., and Fischer, H.-M. 2015. Stable Fluorescent and Enzymatic Tagging of Bradyrhizobium diazoefficiens to Analyze Host-Plant Infection and Colonization. Mol. Plant. Microbe. Interact. 28:959–967

Lee, C. K., de Anda, J., Baker, A. E., Bennett, R. R., Luo, Y., Lee, E. Y., Keefe, J. A., Helali, J. S., Ma, J., Zhao, K., Golestanian, R., O'Toole, G. A., and Wong, G. C. L. 2018. Multigenerational memory and adaptive adhesion in early bacterial biofilm communities. Proc. Natl. Acad. Sci. U. S. A. 115:4471–4476

Lei, Z.-Y., Chen, J.-J., Cao, Z.-J., Ao, M.-Z., and Yu, L.-J. 2018. Efficacy of Aeschynomene indica Linn leaves for wound healing and isolation of active constituent. J. Ethnopharmacol.

Nagy, A., Mowery, J., Bauchan, G. R., Wang, L., Nichols-Russell, L., and Nou, X. 2015. Role of Extracellular Structures of Escherichia coli O157:H7 in Initial Attachment to Biotic and Abiotic Surfaces. Appl. Environ. Microbiol. 81:4720–7

Neubauer, C., Dalleska, N. F., Cowley, E. S., Shikuma, N. J., Wu, C.-H., Sessions, A. L., and Newman, D. K. 2015. Lipid remodeling in Rhodopseudomonas palustris TIE-1 upon loss of hopanoids and hopanoid methylation. Geobiology. 13:443–453

Ochs, D., Kaletta, C., Entian, K., Beck-sickinger, A., Porallal, K., Chemie, O., Tubingen, U., and Institut, B. 1992. Cloning, Expression, and Sequencing of Squalene-Hopene Cyclase,. 174:298–302

Parsons, R., Silvester, W. B., Harris, S., Gruijters, W. T., and Bullivant, S. 1987. Frankia vesicles provide inducible and absolute oxygen protection for nitrogenase. Plant Physiol. 83:728–731

Persat, A. 2017. Bacterial mechanotransduction. Curr. Opin. Microbiol. 36:1–6

Petrova, O. E., and Sauer, K. 2012. Sticky situations: Key components that control bacterial surface attachment. J. Bacteriol. 194:2413–2425

Pini, F., East, A. K., Appia-Ayme, C., Tomek, J., Karunakaran, R., Mendoza-Suarez, M., Edwards, A., Terpolilli, J. J., Roworth, J., Downie, J. A., and Poole, P. S. 2017. Bacterial Biosensors for in Vivo Spatiotemporal Mapping of Root Secretion. Plant Physiol. 174:1289–1306

Regensburger, B., and Hennecke, H. 1983. RNA polymerase from Rhizobium japonicum. Arch. Microbiol. 135:103–109

Ricci, J. N., Coleman, M. L., Welander, P. V., Sessions, A. L., Summons, R. E., Spear, J. R., and Newman, D. K. 2014. Diverse capacity for 2-methylhopanoid production correlates with a specific ecological niche. ISME J. 8:675–684

Richards, F. J. 1959. NA Flexible Growth Function for Empirical Useo Title. J. Exp. Bot. 10:290–301

Saenz, J. P., Grosser, D., Bradley, A. S., Lagny, T. J., Lavrynenko, O., Broda, M., and Simons, K. 2015. Hopanoids as functional analogues of cholesterol in bacterial membranes. Proc. Natl. Acad. Sci. U. S. A. 112:11971–11976

Schindelin, J., Arganda-Carreras, I., Frise, E., Kaynig, V., Longair, M., Pietzsch, T., Preibisch, S., Rueden, C., Saalfeld, S., Schmid, B., Tinevez, J. Y., White, D. J., Hartenstein, V., Eliceiri, K., Tomancak, P., and Cardona, A. 2012. Fiji: An open-source platform for biological-image analysis. Nat. Methods. 9:676–682

Schmerk, C. L., Welander, P. V, Hamad, M. A., Bain, K. L., Bernards, M. A., Summons, R. E., and Valvano, M. A. 2015. Elucidation of the Burkholderia cenocepacia hopanoid biosynthesis pathway uncovers functions for conserved proteins in hopanoid-producing bacteria. Environ. Microbiol. 17:735–750

Schneider, C. A., Rasband, W. S., and Eliceiri, K. W. 2012. *NIH Image to ImageJ: 25 years of Image Analysis HHS Public Access.*

Silipo, A., Vitiello, G., Gully, D., Sturiale, L., Chaintreuil, C., Fardoux, J., Gargani, D., Lee, H.-I., Kulkarni, G., Busset, N., Marchetti, R., Palmigiano, A., Moll, H., Engel, R., Lanzetta, R., Paduano, L., Parrilli, M., Chang, W.-S., Holst, O., Newman, D. K., Garozzo, D., D'Errico, G., Giraud, E., and Molinaro, A. 2014. Covalently linked hopanoid-lipid A improves outer-membrane resistance of a Bradyrhizobium symbiont of legumes. Nat. Commun. 5:5106

Siryaporn, A., Kuchma, S. L., O'Toole, G. A., and Gitai, Z. 2014. Surface attachment induces Pseudomonas aeruginosa virulence. Proc. Natl. Acad. Sci. U. S. A. 111:16860–5

Somado, E. A., Becker, M., Kuehne, R. F., Sahrawat, K. L., and Vlek, P. L. G. 2003. Combined Effects of Legumes with Rock Phosphorus on Rice in West Africa. Agron. J. 95:1172

Sugiyama, A., and Yazaki, K. Root Exudates of Legume Plants and Their Involvement in Interactions with Soil Microbes.

Swapna M.M, S., R., P., K.P., A., C.N, M., and N.P, R. 2011. A review on the medicinal and edible aspects of aquatic and wetland plants of India. J. Med. Plants Res. 5:7163–7176

Syren, P.-O., Henche, S., Eichler, A., Nestl, B. M., and Hauer, B. 2016. Squalene-hopene cyclases-evolution, dynamics and catalytic scope. Curr. Opin. Struct. Biol. 41:73–82

Szparaga, A., and Kocira, S. 2018. Generalized logistic functions in modelling emergence of Brassica napus L. PLoS One. 13:1–14

Vilcheze, C., Llopiz, P., Neunlist, S., Poralla, K., and Rohmer, M. 1994. Prokaryotic triterpenoids: New hopanoids from the nitrogen-fixing bacteria Azotobacter vinelandii, Beijerinckia indica and Beijerinckia mobilis. Microbiology. 140:2749–2753

Welander, P. V, Coleman, M. L., Sessions, A. L., Summons, R. E., and Newman, D. K. 2010. Identification of a methylase required for 2-methylhopanoid production and implications for the interpretation of sedimentary hopanes. Proc. Natl. Acad. Sci. U. S. A. 107:8537–8542

Welander, P. V, Doughty, D. M., Wu, C., Mehay, S., and Roger, E. 2012. Identification and characterization of Rhodopseudomonas palustris TIE-1 hopanoid biosynthesis mutants. 10:163–177

Wu, C.-H., Bialecka-Fornal, M., and Newman, D. K. 2015. Methylation at the C-2 position of hopanoids increases rigidity in native bacterial membranes. Elife. 4

Supplemental Material

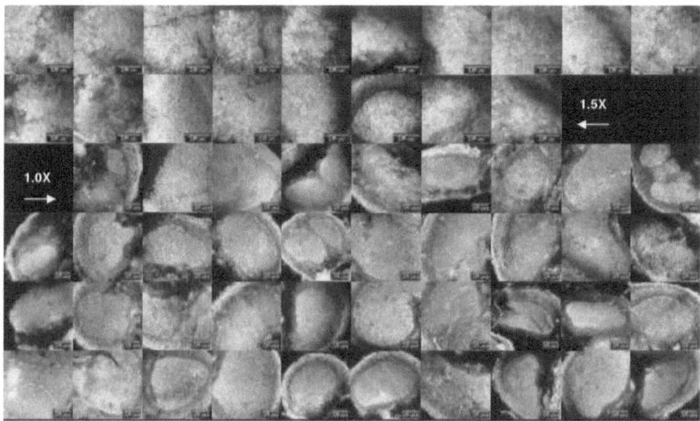

Figure S1. Confocal images of cross-sections of wild-type-infected *A. afraspera* nodules at 24 dpi illustrating plant cell walls (Calcofluor, cyan), live bacteria (SYTO9, yellow), and membrane-compromised bacteria and plant nuclei (propidium iodide, magenta). Nodules were collected from 3 plants.

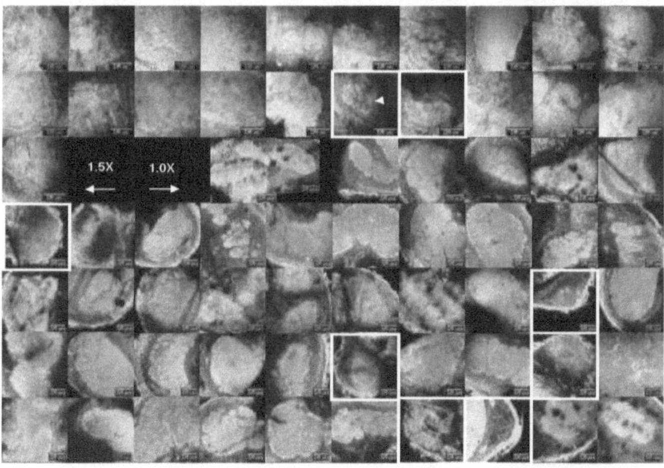

Figure S2. Confocal images of cross-sections of Δ*hpnH*-infected *A. afraspera* nodules at 24 dpi illustrating plant cell walls (Calcofluor, cyan), live bacteria (SYTO9, yellow), and dead bacteria and plant nuclei (propidium iodide, magenta). Nodules were collected from 3 plants. White boxes highlight small nodules. White arrow indicates a likely plant defense reaction.

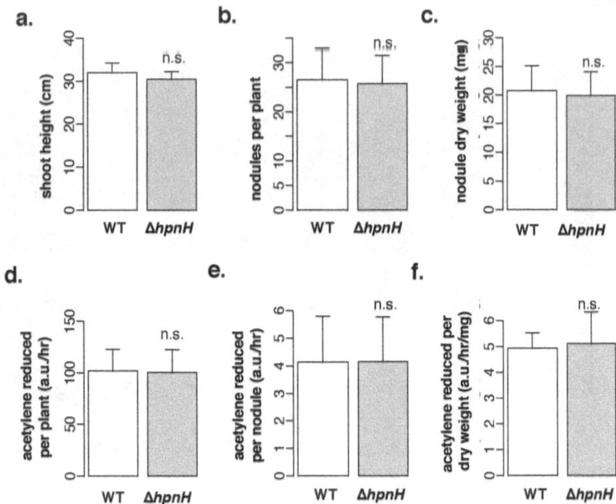

Figure S3. Average (a) shoot height, (b) nodules per plant, (c) nodule dry weight per plant, (d) acetylene reduction per plant, (e) acetylene reduction per nodule, and (f) acetylene reduction per nodule dry weight for *A. asfrapera* inoculated with wild type or Δ*hpnH* at 40 dpi. N=4 plants per bar; error bars represent one standard deviation. Results of two-tailed t-tests between wild type and Δ*hpnH* are denoted as follows: n.s., p>0.05.

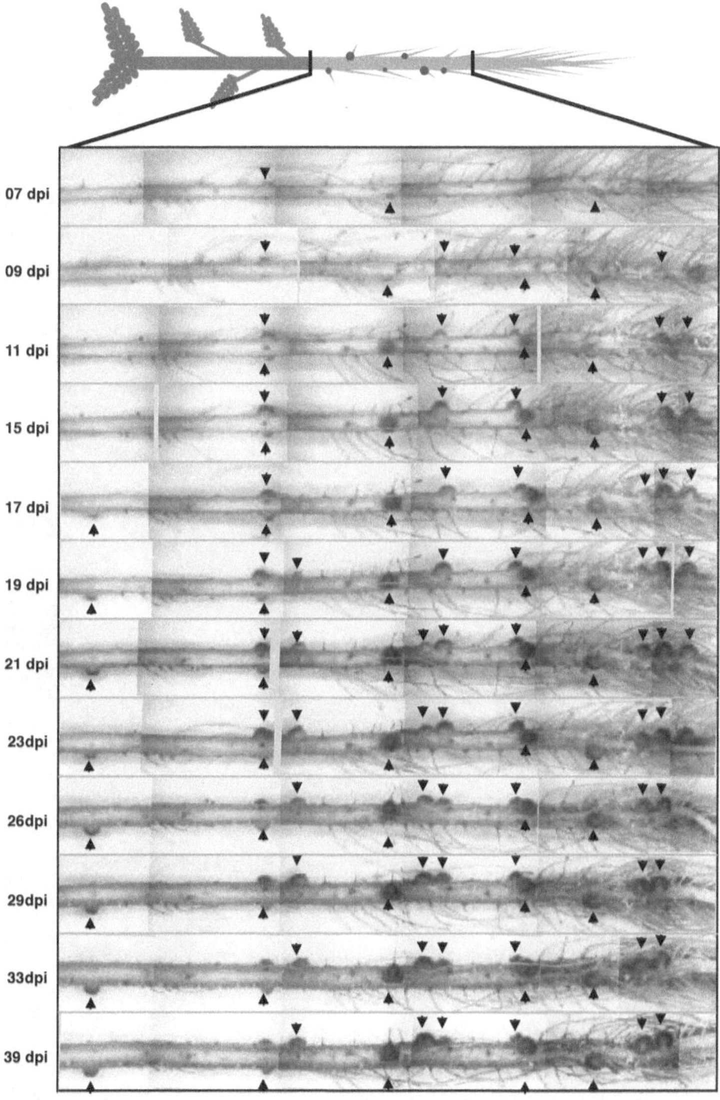

Figure S4. Reconstructed images of the root system of a wild-type-infected *A. afraspera* plant. Nodules fully visible in at least five time points are indicated with black arrowheads.

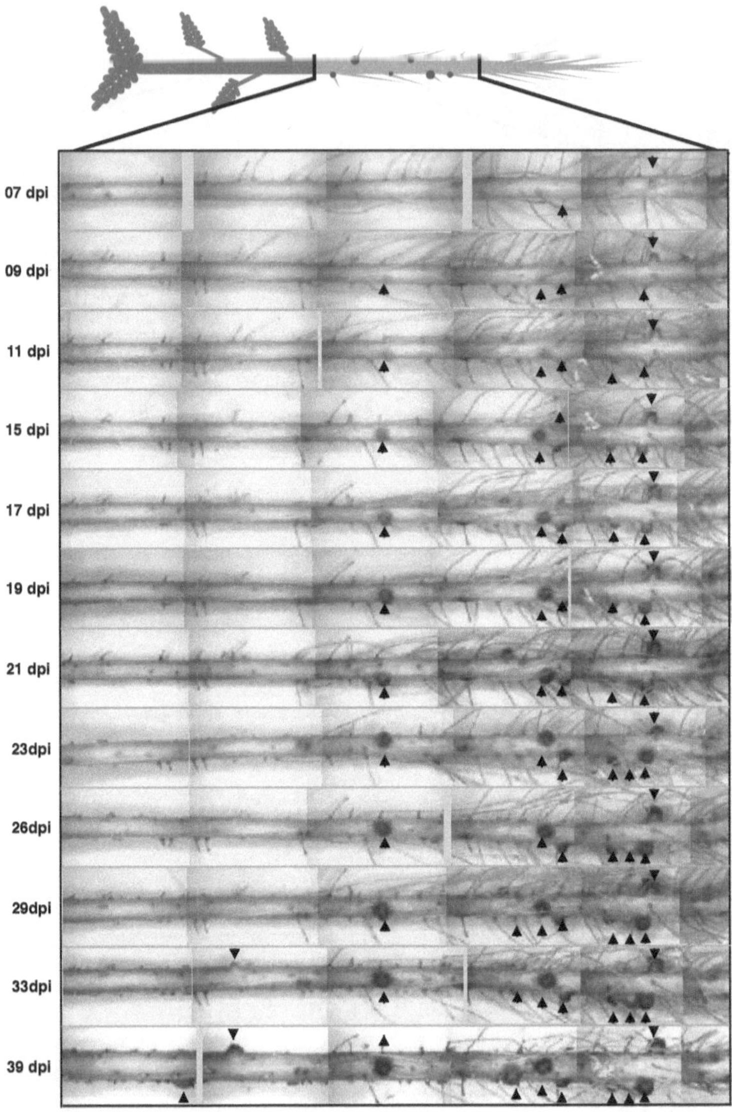

Figure S5. Reconstructed images of the root system of a Δ*hpnH*-infected *A. afraspera* plant. Nodules fully visible in at least five time points are indicated with black arrowheads.

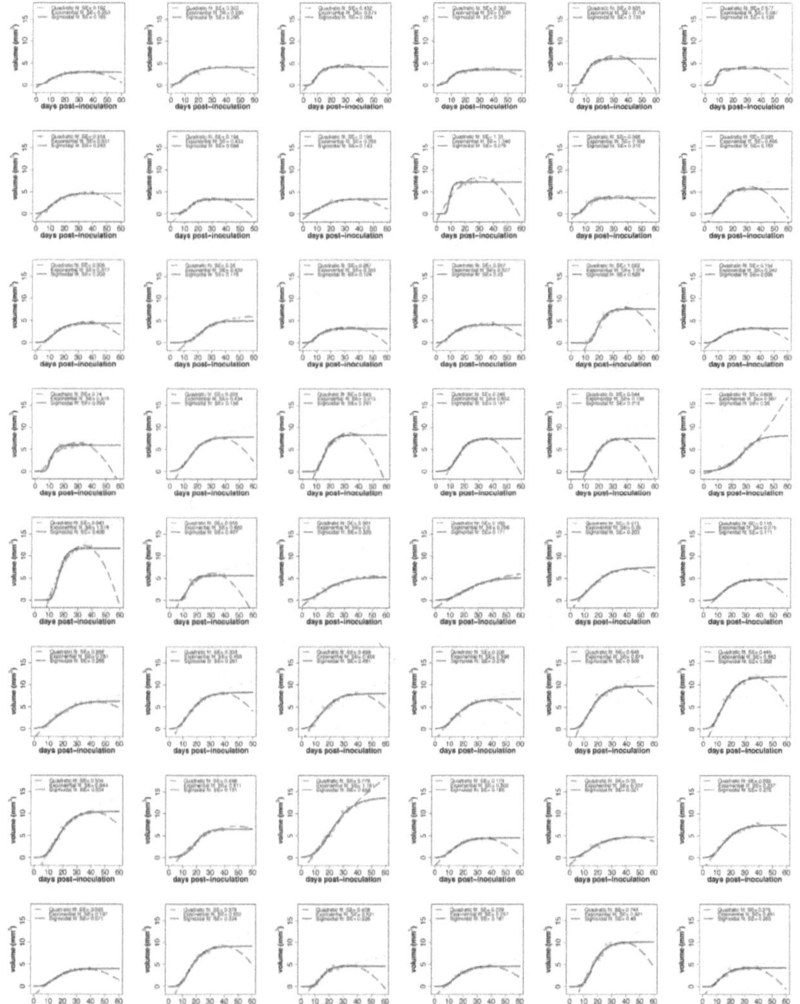

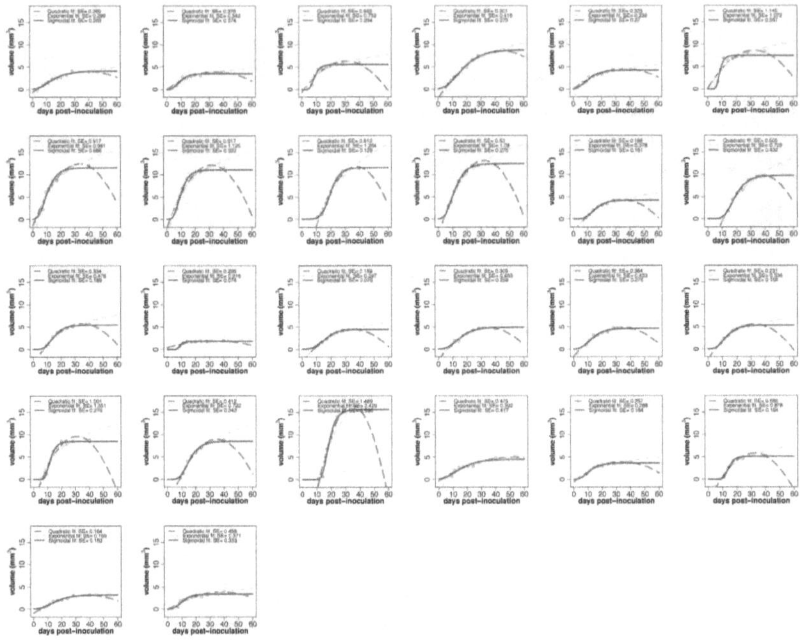

Figure S6. Nodule growth plots for all 74 wild-type-infected nodules fit with quadratic (orange; long dashed lines), exponential (yellow; short dashed lines), or sigmoidal (blue; solid lines) models. Standard errors (SE) for each model are shown.

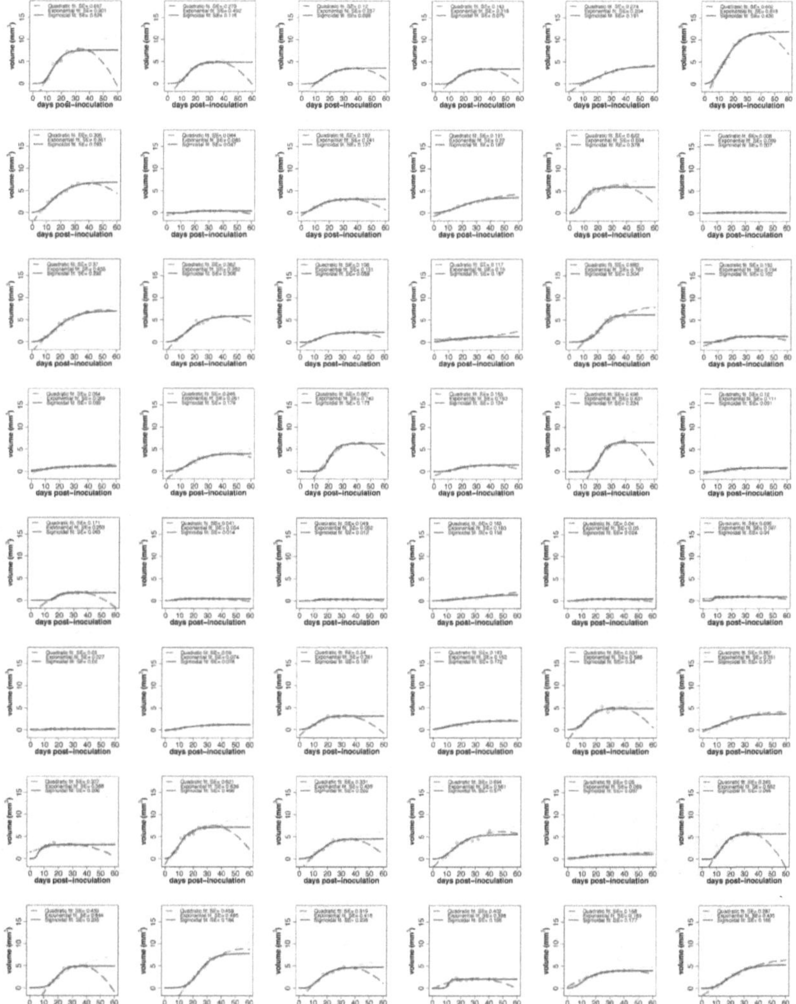

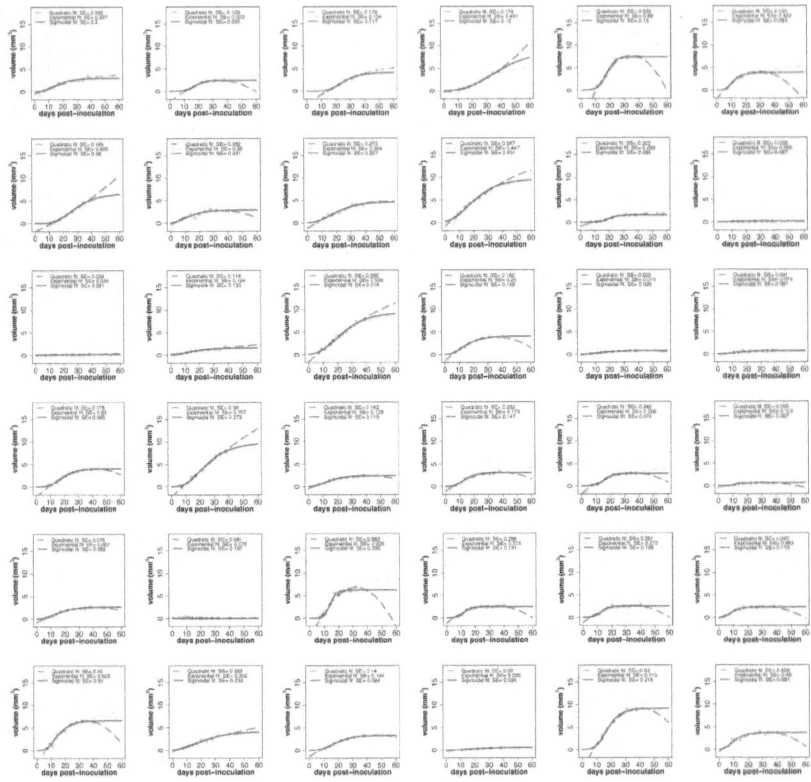

Figure S7. Nodule growth plots for all 84 Δ*hpnH*-infected nodules fit with quadratic (orange; long dashed lines), exponential (yellow; short dashed lines), or sigmoidal (blue; solid lines) models. Standard errors (SE) for each model are shown.

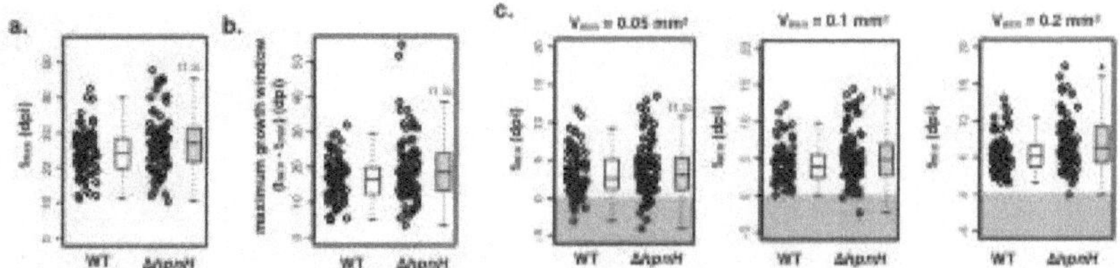

Figure S8. (a) Jitter and box plots of **tmax** values for all wild-type- and Δ*hpnH*-infected nodules. (b) Jitter and box plots of maximum growth windows for all wild-type- and Δ*hpnH*-infected nodules. (c) Jitter and box plots of **tmin** values (as determined by extrapolation using sigmoidal fits of nodule growth curves) for all wild-type- and Δ*hpnH*-infected nodules, in which **Vmin** is defined as 0.05 mm^3, 0.1 mm^3, 0.2 mm^3. Green shading highlights negative **tmin** values. Results of KS-tests between wild-type and Δ*hpnH* nodules are denoted as follows: *, p<0.05; n.s., p>0.05.

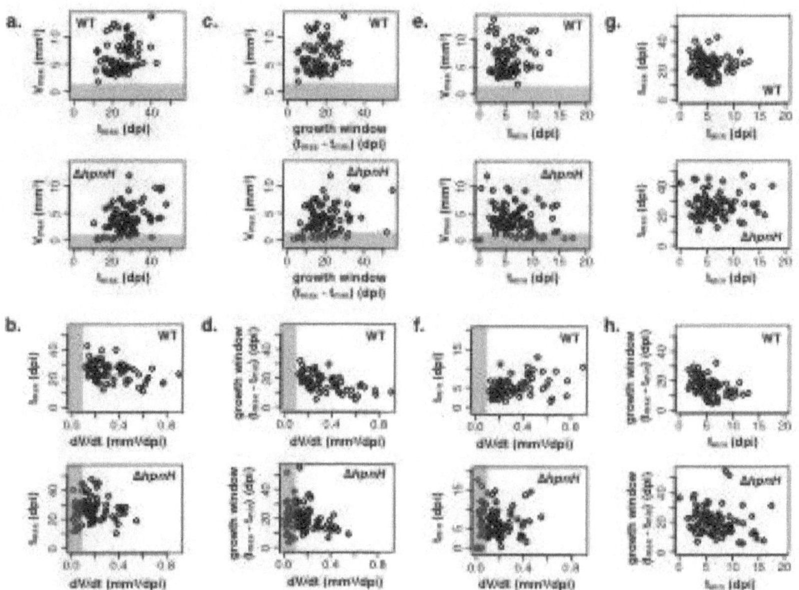

Figure S9. (a-b) Scatter plots of **tmax** vs. (a) **dV/dt** and (b) **Vmax** for all wild-type- (open circles) and Δ*hpnH*- (grey circles) infected nodules. Green regions highlight values below what is observed for wild type. (c-d) Scatter plots of maximum growth windows vs. (c) **dV/dt** and (d) **Vmax**. (e-f) Scatter plots of **tmin** vs. (c) **dV/dt** and (d) **Vmax**. (g-h) Scatter plots of **tmin** vs. (a) **tmax** and (b) maximum growth windows.

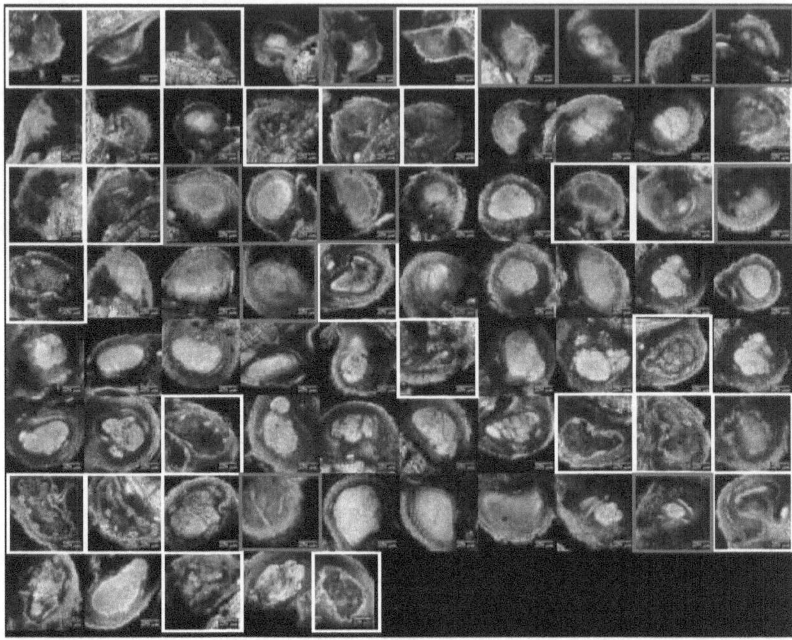

Figure S10. Confocal sections of small (<0.5 mm radius) Δ*hpnH*-infected nodules harvested at 40 dpi. Sections were stained with Calcofluor (cyan), SYTO9 (yellow), and propidium iodide (magenta). N=74 nodules harvested from 5 plants. White boxes highlight under-infected nodules. Magenta boxes indicate nodules primarily containing membrane-compromised cells.

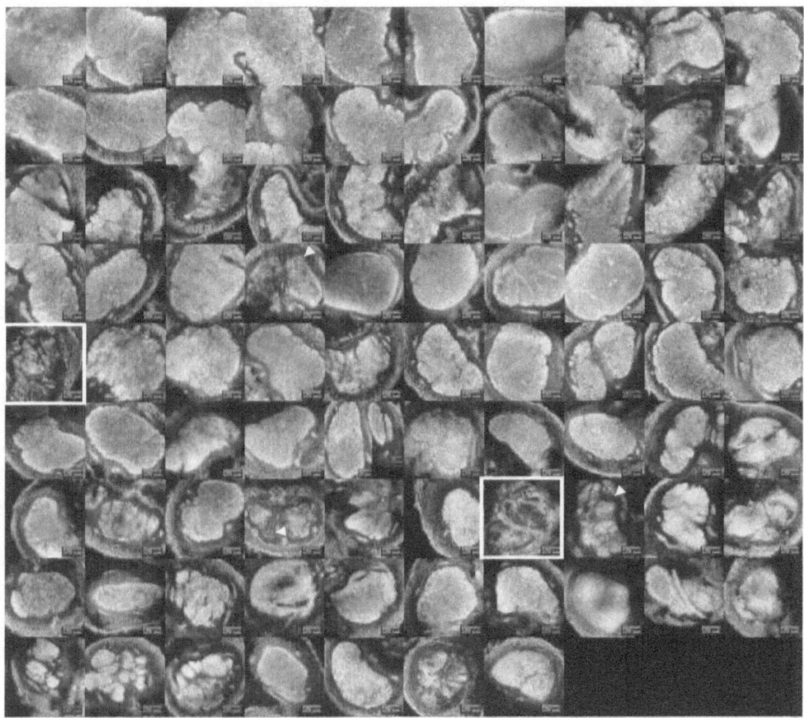

Figure S11. Confocal sections of large (>0.5 mm radius) ΔhpnH-infected nodules harvested at 40 dpi. Sections were stained with Calcofluor (cyan), SYTO9 (yellow), and propidium iodide (magenta). N=87 nodules harvested from 5 plants. White boxes highlight under-infected nodules.

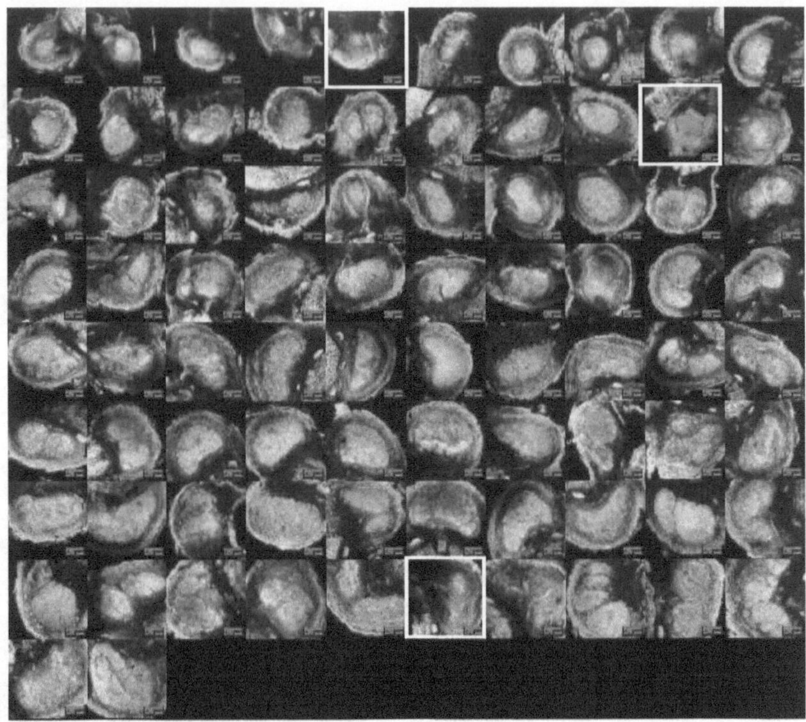

Figure S12. Confocal sections of small (<0.5 mm radius) wild type-infected nodules harvested at 10 dpi. Sections were stained with Calcofluor (cyan), SYTO9 (yellow), and propidium iodide (magenta). N=80 nodules harvested from 5 plants. White boxes highlight under-infected nodules.

Figure S13. Confocal sections of small (<0.5 mm radius) wild type-infected nodules harvested at 25 dpi. Sections were stained with Calcofluor (cyan), SYTO9 (yellow), and propidium iodide (magenta). N=82 nodules harvested from 5 plants. Magenta boxes indicate nodules primarily containing membrane-compromised cells.

Figure S14. Confocal sections of wild type-infected nodules harvested at 40 dpi. Sections were stained with Calcofluor (cyan), SYTO9 (yellow), and propidium iodide (magenta). N=117 nodules harvested from 5 plants. Magenta boxes indicate nodules primarily containing membrane-compromised cells.

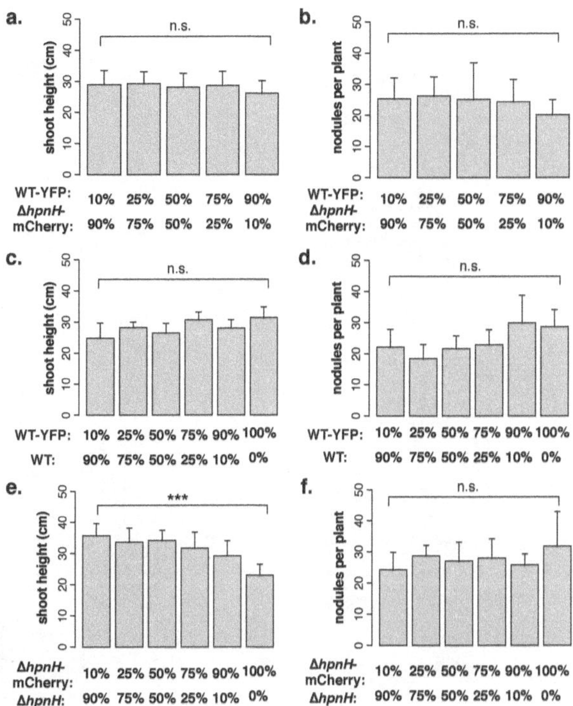

Figure S15. Average shoot height (a) and number of nodules (b) for plants co-inoculated with Δ*hpnH*-mCherry and WT-YFP strains, recorded at 45 dpi. Average shoot height (c) and number of nodules (d) for plants co-inoculated with WT and WT-YFP strains, recorded at 40 dpi. Average shoot height (e) and number of nodules (f) for plants co-inoculated with Δ*hpnH* and Δ*hpnH*-mCherry strains, recorded at 50 dpi. N=7-8 plants per bar for all panels. Error bars represent one standard deviation. Results of two-tailed t-tests are denoted as follows: n.s., p>0.05; ***, p<0.0001.

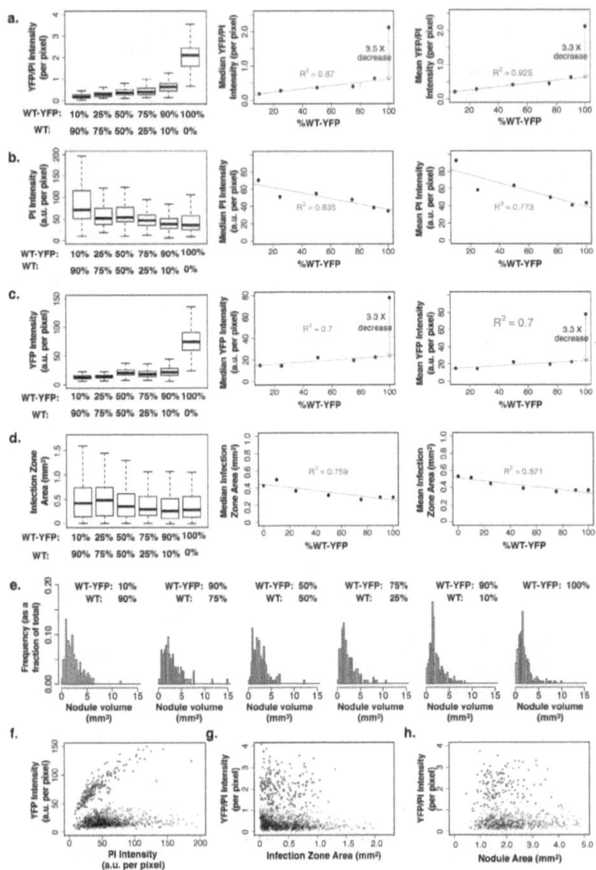

Figure S16. (a-d) Intensity ratio of YFP to mCherry (a), mCherry intensity (b), and YFP intensity (c) per pixel within infection zones of nodules co-inoculated with Δ*hpnH*-mCherry and WT-YFP strains. (d) Cross-sectional area of infection zones of nodules co-inoculated with Δ*hpnH*-mCherry and WT-YFP strains. For (a-d), N=132, 125, 143, 143 and 110 nodules for 10%, 25%, 50%, 75% and 90% WT-YFP strain mixtures, respectively, which were sectioned and fixed between 45-50 dpi. (e) Nodule volume distributions from plants co-inoculated with Δ*hpnH*-mCherry and WT-YFP strains at 45 dpi. Sample sizes are N = 251, 200, 227, 204, and 149 nodules pooled from N = 8, 7, 7, 8, and 7 plants for the 10%, 25%, 50%, 75% and 90% WT-YFP strain mixtures, respectively. (f) Scatter plots of mCherry vs. YFP intensities per pixel within infection zones of nodules co-inoculated with Δ*hpnH*-mCherry and WT-YFP strains. (g-h) Scatter plots of YFP/mCherry intensity ratios per pixel in infection zones vs. infection zone (g) and nodule (h) cross-section areas for nodules co-inoculated with Δ*hpnH*-mCherry and WT-YFP. Scatter plots contain data pooled from all ratios.

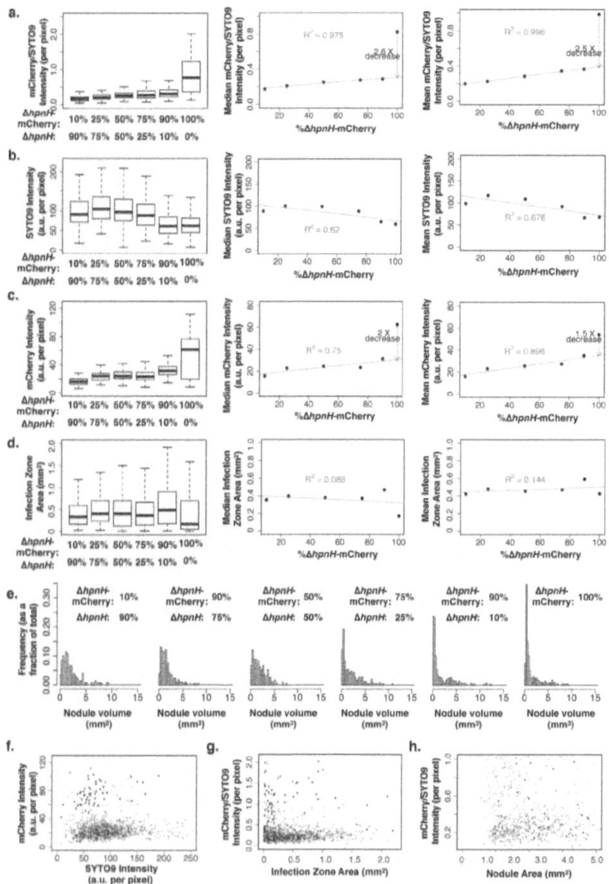

Figure S17. (a-d) Intensity ratio of YFP to propidium iodide (PI) (a), PI intensity (b), and YFP intensity (c) per pixel within infection zones of nodules co-inoculated with WT and WT-YFP strains. (d) Cross-sectional area of infection zones of nodules co-inoculated with WT and WT-YFP strains. For (a-d), N = 141, 95, 134, 147, 133, and 167 nodules for 10%, 25%, 50%, 75%, 90% and 100% WT-YFP strain mixtures, respectively, which were sectioned and fixed between 40-45 dpi. (e) Nodule volume distributions from plants co-inoculated with WT and WT-YFP strains at 40 dpi. Sample sizes are N = 183, 116, 161, 172, 232, and 248 nodules pooled from N = 8, 7, 8, 8, 8, and 8 plants for the 10%, 25%, 50%, 75% and 90% WT-YFP strain mixtures, respectively. (f) Scatter plots of PI vs. YFP intensities per pixel within infection zones of nodules co-inoculated with WT and WT-YFP strains. (g-h) Scatter plots of YFP/PI intensity ratios per pixel in infection zones vs. infection zone (g) and nodule (h) cross-section areas for nodules co-inoculated with WT and WT-YFP strains. Scatter plots contain data pooled from all ratios.

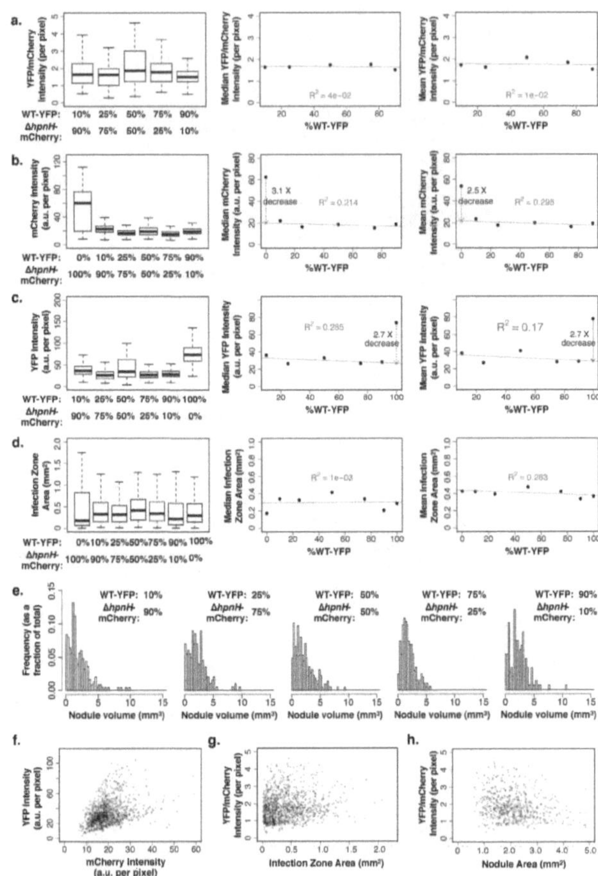

Figure S18. (a-d) Intensity ratio of mCherry to SYTO9 (a), SYTO9 intensity (b), and mCherry intensity (c) per pixel within infection zones of nodules co-inoculated with Δ*hpnH*-mCherry and Δ*hpnH* strains. (d) Cross-sectional area of infection zones of nodules co-inoculated with Δ*hpnH*-mCherry and Δ*hpnH* strains. For (a-d), N = 117, 107, 128, 137, 103 and 50 nodules for 10%, 25%, 50%, 75%, 90% and 100% Δ*hpnH*- mCherry strain mixtures, respectively, which were sectioned and fixed between 50-55 dpi. (e) Nodule volume distributions from plants co-inoculated with Δ*hpnH*-mCherry and Δ*hpnH* strains at 45 dpi. Sample sizes are N = 150, 222, 191, 254, 297, and 236 nodules pooled from N = 7, 7, 7, 8, 8, and 8 plants for the 10%, 25%, 50%, 75% and 90% WT-YFP strain mixtures, respectively. (f) Scatter plots of mCherry vs. SYTO9 intensities per pixel within infection zones of nodules co-inoculated with Δ*hpnH*-mCherry and Δ*hpnH* strains. (g-h) Scatter plots of mCherry/SYTO9 intensity ratios per pixel in infection zones vs. infection zone (g) and nodule (h) cross-section areas for nodules co-inoculated with Δ*hpnH*-mCherry and Δ*hpnH* strains. Scatter plots contain data pooled from all strain ratios.

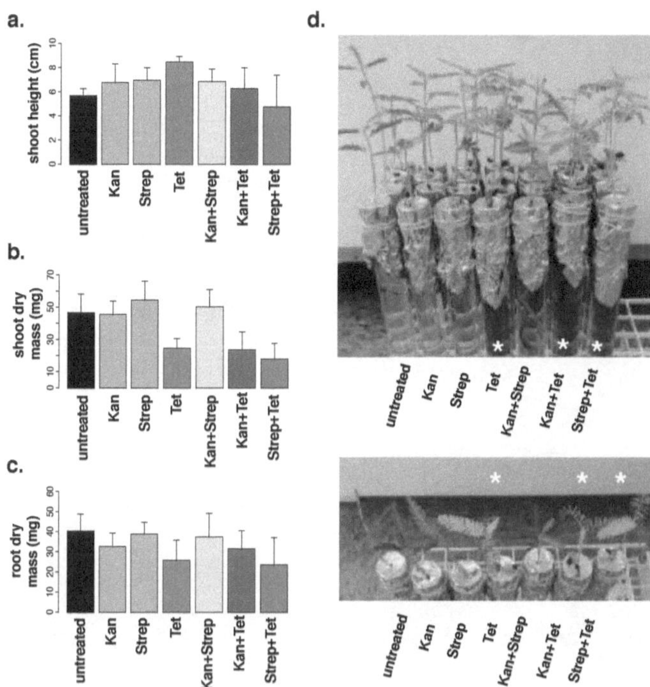

Figure S19. Average (a) shoot height, (b) shoot dry mass and (c) root dry mass for non-inoculated *A. afraspera* plants grown in BNM supplemented with kanamycin, streptomycin, or tetracycline for 2 weeks under normal growth conditions. N=4 plants per condition; error bars represent one standard deviation. (d-e) Images of *A. afraspera* plants after 2 weeks of antibiotic treatment. Asterisks indicate plants grown in tetracycline-supplemented medium.

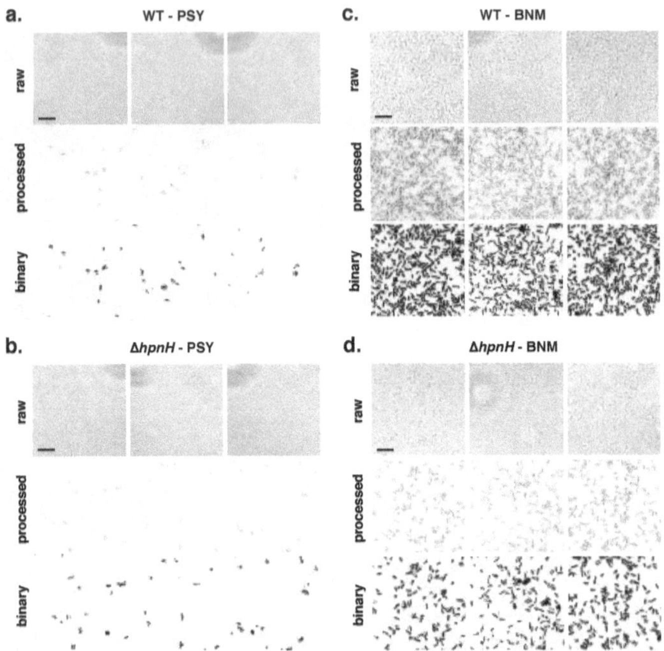

Figure S20. Surface attachment of wild type (a,c) and Δ*hpnH* (b,d) incubated on glass coverslips in different media. For each panel, raw phase images (top row), background-subtracted images (middle row), and binary images with cells shown in black (bottom row) are shown. Scale bars represent 5 μm.

Table S1. Motile cell counts and mean swimming speeds for wild-type and Δ*hpnH* *B. diazoefficiens*.

Movie S1. Sample 15 second excerpt from a 5-minute recording of *B.diazoefficiens* WT in PSY medium. Centroid trajectories of sample swimming cells are overplayed as a yellow trace. Recording has been sped up at 2x the acquisition speed.

Movie S2. Sample 15 second excerpt from a 5-minute recording of *B.diazoefficiens* Δ*hpnH* in PSY medium. Centroid trajectories of sample swimming cells are overplayed as a yellow trace. Recording has been sped up at 2x the acquisition speed.

Movie S3. Sample 15 second excerpt from a 5-minute recording of *B.diazoefficiens* WT in BNM medium. Centroid trajectories of sample swimming cells are overplayed as a yellow trace. Recording has been sped up at 2x the acquisition speed.

Movie S4. Sample 15 second excerpt from a 5-minute recording of *B.diazoefficiens ΔhpnH* in BNM medium. Centroid trajectories of sample swimming cells are overplayed as a yellow trace. Recording has been sped up at 2x the acquisition speed.

Chapter 4

HOPANOIDS CONFER ROBUSTNESS TO PHYSICOCHEMICAL VARIABILITY IN THE NICHE OF THE PLANT SYMBIONT *BRADYRHIZOBIUM DIAZOEFFICIENS*

Introduction

The soil is a precious ecosystem. The health of the soil–measured by organic matter and nutrient content, moisture retention, and the microbial community–predicts how well plants will grow (1). While practices to maintain healthy soil have been known for centuries, as agriculture faces the threat of climate change, these sustainable land management practices have garnered new attention as potential climate change mitigation strategies (2, 3). Crop rotation is an ancient land management strategy that restores nutrients to the soil. Legumes, such as soybean, peanut, and alfalfa, are an important component of crop rotations because they increase soil nitrogen content, reducing reliance on synthesized nitrogen fertilizers (4). However, legumes cannot do this alone: they rely on a symbiosis with a group of polyphyletic soil bacteria called rhizobia (5, 6). This symbiosis is a very close interaction, with the bacteria living intracellularly in specialized *de novo* organs called root nodules. Low pH, low oxygen, and elevated osmolarity are maintained within the nodule environment (7–9). While in some ways stressful for the bacteria, this environment favors bacterial conversion of nitrogen gas to bioavailable ammonia which eventually is exchanged for reduced carbon in the form of dicarboxylic acids.

One way to improve legume use in crop rotations as a sustainable nitrogen fertilizer is to improve the efficiency of the legume-rhizobia symbiosis. Rhizobia strains with greater symbiotic efficiency, as measured by nitrogen fixation rates and legume growth, have been isolated and applied to the soil or to legume seeds. This strategy has been used successfully at scale with legume crops such as soybean in Brazil (10–12). However, the rhizobia often fail to stably integrate into the soil community, so these inoculations must be repeated each year (13–16). Beyond the symbiosis itself, the rhizobia can have positive effects on the plant when living in the soil, such as relieving salinity stress and increasing water and nutrient uptake (17–19). These positive effects will become even more important as the climate changes, leading to drastic changes in precipitation and thus soil water potential and osmotic strength. To successfully use legumes in crop rotations to increase soil nitrogen content as the climate changes, rhizobia must be successful in both the symbiosis and surviving the soil environment. The fact that these bacteria are ubiquitous indicates that they have strategies to accomplish both goals (20).

As recently discussed, possession of an adaptable outer membrane may provide an important fitness advantage in this context (9). Rhizobia produce modified lipid A, a major component of the lipopolysaccharides (LPS) that make up the outer leaflet of the outer membrane (21, 22). These modifications trend toward increasing the hydrophobicity and other cohesive interactions that lead to a more robust outer membrane (23, 24). Additionally, a subset of rhizobia, mostly within the *Bradyrhizobia* clade, make hopanoids, a class of sterol-like lipids, which maintain resistance to environmental stressors such as pH, temperature, antibiotics and foster successful symbioses (**Figure 1**) (25, 26). Some *Bradyrhizobia* also attach hopanoids to lipid A, which appears to act as a hydrophobic hook into the inner leaflet of the outer membrane (27–29). This hopanoid attached to lipid A (HoLA) is synthesized from extended hopanoids, a subclass of hopanoids with an added hydrophilic tail (25).

Intriguingly, generating a mutant that is unable to make any type of hopanoid by removing the first committed step in hopanoid biosynthsis (Δshc) has evaded realization in *Bradyrhizobium diazoefficiens*, pointing to an essential role for hopanoids in this strain (25). Removing the ability to synthesize C_{35} or "extended" hopanoids ($\Delta hpnH$), however, was achieved, and has a large effect on the fitness of *B. diazoefficiens in vitro* (25, 30). The $\Delta hpnH$ mutant manifests growth defects at high osmolarity and is unable to grow under low pH or microaerobic conditions–all conditions thought to characterize the root nodule. While the $\Delta hpnH$ mutant exhibits defects *in planta*, especially in root nodule initiation, the nitrogen fixation rate in symbiosis with the tropical legume *Aeschynomene afraspera* when normalized for nodule dry weight is not significantly different between the Δ*hpnH* mutant and WT and the majority of Δ*hpnH*-infected nodules grow at rates comparable to the WT (30). Given the growth defects of the Δ*hpnH* mutant observed *in vitro* in the presence of environmental stresses expected within the root nodule, these results were surprising. Here, we investigate this paradox by exploring the nuanced interplay between the lack of extended hopanoids and an environmentally relevant concentration range of osmolytes and cations.

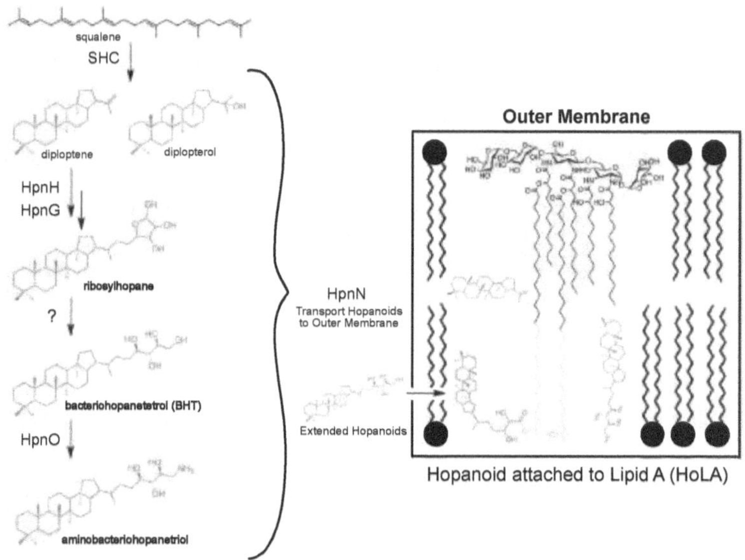

Figure 1. Hopanoid biosynthesis and HoLA structure. The biosynthesis of hopanoids is shown on the left from squalene to the unextended hopanoids (magenta) to the extended hopanoids (green). All of these hopanoids may be methylated at the C-2 position by the hopanoid methylase HpnP. These hopanoids are transferred to the outer membrane by the HpnN hopanoid transporter. Extended hopanoids are attached to the very long chain fatty acid (yellow) on lipid A to create HoLA which also resides in the outer membrane.

Materials and Methods

Bacterial strains, Culture Media, and Chemicals

All strains used in this study are described in Table S1 in the supplemental material. All strains were grown aerobically with shaking at 250 rpm. *Escherichia coli* strains were grown in lysogeny broth (LB) at 37°C (31). *B. diazoefficiens* strains were grown at 30°C in either peptone-salts-yeast extract medium with 0.1% arabinose (PSY) (32, 33) or arabinose-gluconate medium (AG) (34, 35). The pH of the AG medium was adjusted to pH 6.6 using a NaOH solution or to pH 5 using a HCl solution. For pH 5 AG media, the HEPES buffer component was replaced with 5.5 mM MES buffer for a total of 11 mM MES. A low divalent cations pH 5 AG media was made containing 45 μM $CaCl_2$ and 400 μM $MgSO_4$. Inositol, sorbitol, $CaCl_2$, or $MgCl_2$ were added to the appropriate base medium. For induction of the cumate inducible promoter, cumate was added to liquid or solid medium for a final concentration of 25 μM from a 400x stock solution in ethanol (36–39). Agar plates were made containing 1.5% (w/v) agar. Antibiotics were used for selection at these concentrations (μg/ml): spectinomycin (Sp), 100; kanamycin (Km), 100; tetracycline (Tc), 20 for liquid cultures of *B. diazoefficiens* and 50 for plates for *B. diazoefficiens* and for *E. coli*. Chemicals were purchased from Sigma-Aldrich unless otherwise noted: glycerol (VWR), Hepes (Gold BioTechnology), sodium chloride yeast extract, magnesium sulfate, magnesium chloride, sorbitol, sodium hydroxide (Fisher Scientific), arabinose (Chem-Impex International, Inc), sodium sulfate (Mallinckrodt Chemical), peptone, and agar (Becton Dickinson).

DNA methods, plasmid construction, and transformation

All plasmid constructions and primers used in this study are described in **Table S1** in the supplemental material. Standard methods were used for plasmid DNA isolation and manipulation in *E. coli* (40). The strong constitutive promoter P$_{rrn\text{-}mut2}$ (41) was annealed from oligonucleotides Prrn-mut2_f and Prrn-mut2_r and cloned into *Hpa*I/*Bsr*GI-digested pQH2, resulting in plasmid pQH2-Prrn-mut2. The resulting inducible system (containing P$_{bla\text{-}mut1T}$-*cymR** and P$_{rrn\text{-}mut2}$ flanked by by *cuO*) was subsequently excised with *Spe*I and *Pci*I and ligated into pRJPaph-lacZYA prepared with *Spe*I and *Nco*I, resulting in plasmid pRJPcu1-lacZYA. The PCR product of the *shc* gene was cloned into the pRJPcu plasmids respectively to obtain expression plasmids. The pRJPcu-*shc* plasmid was mobilized into WT, followed by the pGK302, the markerless deletion vector to delete *shc* (blr3004) (42). Plasmids were mobilized by conjugation from *E. coli* S17-1 into *B. diazoefficiens* strains as previously described with the following modifications (43). The pRJPcu-*shc* plasmid was stably integrated as a single copy into the *scoI* downstream region of *B. diazoefficiens* as described previously (44).

Induction conditions and reporter activity measurements

Cultures were grown in PSY to a mid-exponential phase and induced with 25 μM (final concentration) cumate or pure ethanol (the solvent for cumate) for controls. For quantitative LacZ assays, cells were centrifuged (5000 x g) and washed twice. β-Galactosidase assay was done as previously described (31). One biological replicate was defined as an independent culture, each replicate was assayed in technical duplicates of which the arithmetic mean was used for final data plotting.

Streaking strains

Liquid cultures were grown from plates to early stationary phase (OD 0.9-1.2 by Beckman Coulter UV-VIS) in AG media. The cultures were spun down and resuspended to OD_{600} 0.5 in fresh media. 10 µL of culture was spotted on each plate and spread using a sterilized spreader.

Osmometer measurements

The osmolarity of PSY and AG media were measured using a Wescor Vapro 5520 vapor pressure osmometer. Before use, the osmometer was calibrated using 100 mM, 300 mM, and 1000 mM OptiMole standards from ELITechGroup.

B. diazoefficiens *pregrowth*

5 mL of fresh media was inoculated with multiple colonies per strain were picked using a sterile stick. After 2-3 days, when the cultures were visibly turbid, these cultures were subcultured into 5 mL of fresh media and allowed to grow to mid-late exponential phase (OD_{600} 0.5-1). The subculturing was repeated, and the second subculture was used as the inoculum for all experiments unless otherwise noted.

Growth curves

The growth curve assays were performed in 96 well tissue culture plates (Genesee Scientific) using a Spark 10M multimode microplate reader (Tecan, Grödig, Austria). Wells were topped with 50 µL autoclaved mineral oil. Optical density absorbance was taken at 600 nm at 30-minute increments at 30°C with continuous linear shaking.

Growth curve parameter estimation

To estimate the maximum specific growth rate (μ_m) and lag time (λ) for the growth curves, the data was fit using an R application that relies on nonlinear least squares to fit nonlinear models to the following Gompertz curve equation (45, 46):

$$OD_{600} = A * exp\left(-exp\left(\frac{\mu_m}{A}(\lambda - t) + 1\right)\right) + C$$

where A is the final OD_{600}, t is the time in hours, and C is an adjustment for initial OD_{600}. The R application can be found at the following github repository: https://github.com/scott-saunders/growth_curve_fitting. The specific version of the growth fitting R application used was retrieved on July 13, 2021 and can be found here: https://github.com/scott-saunders/growth_curve_fitting/blob/master/growth_curve_fitting_ver0.2.Rmd. This is a direct link to the application that can be run locally: https://scott-h-saunders.shinyapps.io/gompertz_fitting_0v2/#section-parameter-estimates.

Viable-cell plate counts

Viable-cell plate counts were performed by serially diluting samples in fresh AG media. Dilutions spanning 6 orders of magnitude were plated on AG agar plates as 10 µl drips. Plates were incubated at 30°C. Colonies were counted after 4 days for WT and *ΔhpnH* complement and after 5 days for *ΔhpnH*.

Fluorescence measurement

Three biological replicates were grown at 30°C and harvested at mid-exponential (OD_{600} = 0.5). Harvested cells were washed 2x and resuspended in media at an OD_{600} of 0.2. Cells were then transferred to black bottom 96-well plates and stained with 80 nM Di-

4 ANEPPDHQ (ThermoFisher, D36802). All spectroscopical measurements were carried out using a SPARK 20 plate reader (Tecan, Grödig, Austria) equipped with a thermostat capable of maintaining the temperature with the accuracy of ± 1°C. The measurements were carried out at 30°C. Upon reaching a temperature, the sample was first incubated for 30 minutes at 150 rpm using internal sample holder to ensure thermal equilibrium. Fluorescence emission was measured in the 'top reading mode' of the setup in the epi-configuration using a 50/50 mirror and two monochromators (for selecting excitation and emission wavelengths). The sample was excited with xenon flash lamp with the excitation monochromator set to 485 nm (20 nm bandwidth. The fluorescence emission was measured at 540 nm and 670 nm wavelength (20 nm bandwidth each).

General polarization

General polarization (GP) was calculated using the following formula:

$$GP = \frac{I_{540} - I_{670}}{I_{540} + I_{670}}$$

where I is the fluorescence emission intensity at given wavelength after a subtraction of the signal measured for the blank suspension.

Membrane rigidity

For whole cell membrane rigidity measurements, as described in (47), AG-grown aerobic cultures of *B. diazoefficiens* strains were washed once with 4-(2-hydroxyethyl)-1-piperazineethanesulfonic acid (HEPES) buffer (50 mM HEPES, 50 mM sodium chloride (NaCl), pH 7.0) and then resuspended in the same to an OD_{600} ~0.2 with 8 µM of the fluorophore diphenyl hexatriene (DPH). Prior to measurement of fluorescence polarization, samples were incubated in a 30°C water bath in dark for 30 min. Fluorescence polarization measurements (Fluorolog, HORIBA Instruments (Edison, NJ)) were taken using the following parameters: ex 358 nm, slit = 3 mm; em 428 nm, slit = 7 mm; integration time = 1s (48). Three biological replicates were measured, each containing 8 technical replicates.

Results

Hopanoids are conditionally essential in B. diazoefficiens

Due to the previous difficulties isolating a *shc* deletion strain in *B. diazoefficiens*, we decided to construct a conditional SHC expression strain in a Δ*shc* background. The conditional expression system employed a cumate repressor system that was modified for *B. diazoefficiens*, where cumate relieves repression of transcription of the gene of interest (**Table S1**). With the cumate conditional SHC expression strain in hand, we tested the growth of the strain on plates made up of different media commonly used to cultivate rhizobia *in vitro* (**Figure 2**). The cumate conditional SHC expression strain was unable to grow on PSY media without cumate present. Therefore, in PSY media, hopanoid production appears to be essential for the growth of *B. diazoefficiens*. However, when we tested for growth on AG medium plates, we observed moderate growth compared to WT without cumate added. This unexpected result indicated that hopanoids are not essential under this condition. With this in mind, we considered the compositional differences between these two media (**Table 1**). While there are many altered components between the two media, two aspects that stood out to us were the differences in osmolarity and the differences in divalent cation concentrations. Osmolarity is thought to be elevated in the root nodule and can span a wide range in the soil environment (9). Divalent cations are known to stabilize the outer membrane through interactions with LPS, especially calcium (23, 49). The medium osmolarity was 16 mM greater and the divalent cation concentration was almost doubled in PSY compared to the AG medium (**Table 1**), suggesting that hopanoids are necessary to withstand certain levels of osmolytes and/or ionic strength in *B. diazoefficens*.

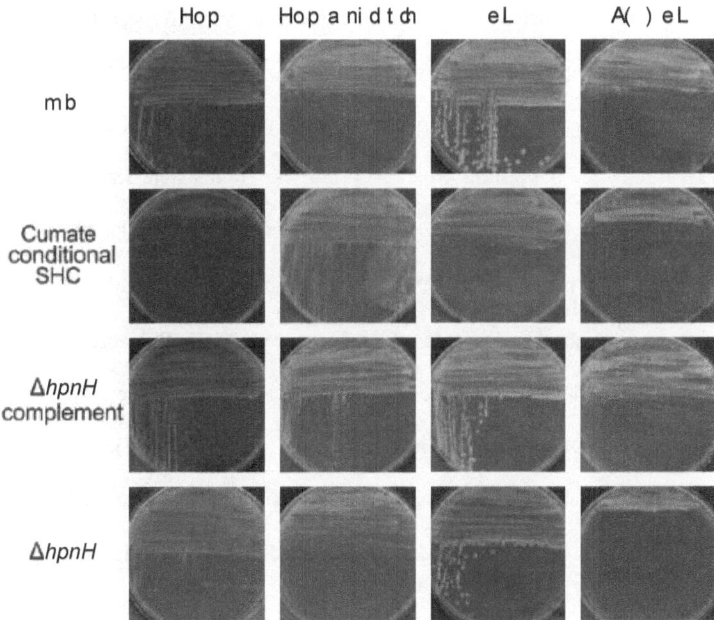

Figure 2. Hopanoids are conditionally essential in *B. diazoefficiens*. A Δ*shc* deletion mutant with a chromosomally integrated cumate-inducible *shc* gene (cumate conditional SHC) is unable to grow on a PSY medium agar plate, but this strain can grow on a PSY agar plate containing cumate which restores hopanoid production. On an AG medium agar plate, the cumate conditional SHC mutant can grow more than on the PSY, but less growth than on PSY with cumate. There is visually greater growth on pH 5 AG medium than on PSY. WT and the Δ*hpnH* complement strains can grow in all conditions. The Δ*hpnH* strain can grow in all conditions as well, with the best growth on the AG medium agar plate.

Despite this realization, we continued to struggle with obtaining a clean *shc* deletion mutant in AG medium, likely due to the fact that the sucrose-selection method we were using to generate the mutant strain (50) exposed it to an osmotic stress beyond the threshold it could tolerate. Accordingly, to probe how the medium composition affects *B. diazoefficiens* strains lacking hopanoids and what this reveals about the ecophysiological role of hopanoids more generally, we turned to the *B. diazoefficiens* Δ*hpnH* strain. To verify that the experimental results using the *B. diazoefficiens* Δ*hpnH* strain are due to the lack of extended hopanoids, we constructed a complement strain with the *hpnH* gene under control of the P_{aphII} constitutive promoter integrated at the *scoI* locus, as used previously to express a range of fluorescent proteins (44). These strains were tested under the same conditions, and we observed that Δ*hpnH* also showed better growth on AG medium plates compared to PSY plates, while the Δ*hpnH* complement strain behaved similarly to WT. Previously, Δ*hpnH* was shown to be unable to grow at pH 6 in PSY medium (25). The pH within the root nodules and in the legume rhizosphere is known to be acidic (7, 51), so we tested the growth of our strains on pH 5 AG media plates (**Figure 1**). Interestingly, while growth was lower for the Δ*hpnH* strain and the *shc* conditional mutant at this pH, they were still able to grow. This result indicates that hopanoids are not essential at low pH under all conditions.

Table 1. Differences between PSY and AG Media

Medium	Buffer	Divalent Cations	Carbon Sources	Osmolarity	pH
PSY	4.3 mM Phosphate	45 µM $CaCl_2$ 400 µM $MgSO_4$	Arabinose Yeast Extract Peptone	53±0.06 mOsM	7
AG	5.5 mM HEPES 5.6 mM MES	90 µM $CaCl_2$ 730 µM $MgSO_4$	Arabinose Yeast Extract Sodium Gluconate	69±1.2 mOsM	6.6

Extended hopanoids protect B. diazoefficiens in stationary phase and at low pH

To better understand how the $\Delta hpnH$ strain grows in AG medium, we quantified different aspects of the growth curve. In PSY, the $\Delta hpnH$ strain has a pronounced defect in both exponential and stationary phase (25). Even when $\Delta hpnH$ and WT cultures were sampled at the same OD_{600} in exponential phase, $\Delta hpnH$ had drastically lower viability than WT; we reasoned that this might be due to initial inoculum viability differences from "overnight" cultures. To test and pre-empt this, we subcultured twice from an initial turbid culture inoculated with colonies from a fresh plate. Using this technique, we observed that the $\Delta hpnH$ mutant strain has only a very slight growth defect compared to WT and the $\Delta hpnH$ complement in the pH 6.6 AG medium (**Figure 3**). To confirm that the similarity in OD_{600} is due to a comparable number of viable cells, we determined colony forming units (CFUs) after 24 hours (mid-late exponential) and 72 hours (stationary phase) of growth. After 24 hours, the CFUs were very similar but after 72 hours, the $\Delta hpnH$ strain was much worse off, leading to a significant difference in CFUs between the WT and $\Delta hpnH$ strain. This result confirms the stationary phase defect observed previously in PSY medium, and that our culturing method successfully removes differences in inocula that influenced previous experiments. Using this approach, we tested the $\Delta hpnH$ strain's growth in the pH 5 AG medium (**Figure 3** and **Figure S1**). Here, we observed a significant defect in growth rate and increased stationary phase death for the $\Delta hpnH$ strain compared to WT and the $\Delta hpnH$ complement. Upon inspection, clumping of the $\Delta hpnH$ strain was observed in the wells of the plate, perhaps indicating death followed by increased biofilm formation.

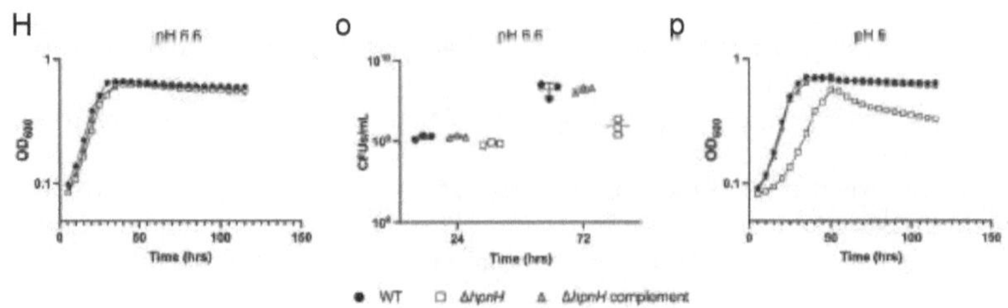

Figure 3. *B. diazoefficiens* Δ*hpnH* strain is sensitive to low pH and stationary phase. (A, C) Growth of WT (black circles), Δ*hpnH* complement (grey triangles), and Δ*hpnH* (white squares) in AG media at pH 6.6 and pH 5 was monitored at an optical density of 600 nm (OD_{600}). Each curve represents the average of three biological replicates. (B) Colony forming units per mL (CFUs/mL) were measured for WT (black circles), Δ*hpnH* complement (grey triangles), and Δ*hpnH* (white squares) strains grown in AG media at pH 6.6 during exponential phase (24 hrs) and stationary phase (72 hrs). Error bars (standard deviation) are included (A-C), but some are smaller than the point markers.

Physicochemical medium conditions affect the growth of B. diazoefficiens ΔhpnH

Having discovered that growth defects are condition-dependent for the hopanoid-deficient Δ*shc* strain, we decided to check whether this was also true for the growth defect of the Δ*hpnH* strain in pH 5 AG medium. First, we tested the effects of osmolarity by adding the nonmetabolizable, nonionic osmolyte inositol (**Figure 4**). The growth rate and lag time parameters were estimated using a Gompertz model. As the concentration of inositol was increased from 25 mM to 400 mM, the growth rate decreased for WT and the Δ*hpnH* complement (**Figure S2**). However, the growth rate for the Δ*hpnH* strain increased to a maximum growth rate with 100 mM inositol added, before decreasing. The Δ*hpnH* strain never reached the same growth rate as WT, but the differences in the growth rate response over this osmolarity range illustrates that the strains experience these conditions very differently and that a "Goldilocks" osmotic zone exists for the Δ*hpnH* mutant where its growth is enhanced. A similar trend was observed with lag time, where the $\Delta hpnH$ strain

exhibited a minimum lag time at 100 mM inositol added. The lag time of the WT and the

$\Delta hpnH$ complement remained relatively constant up to 100 mM inositol added and then

steadily increased. These experiments were also completed with sorbitol as the osmolyte

for WT and the $\Delta hpnH$ strain, showing similar results (**Figure S3**).

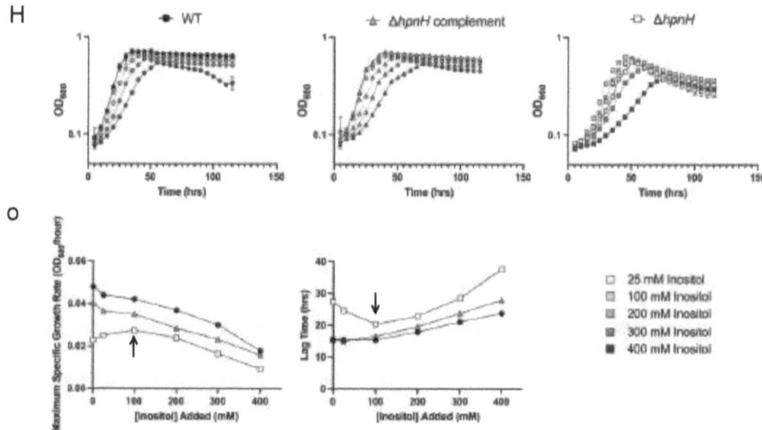

Figure 4. *B. diazoefficiens* $\Delta hpnH$ strain growth is sensitive to the concentration of inositol. (A) Growth of WT (black circles), $\Delta hpnH$ complement (grey triangles), and $\Delta hpnH$ (white squares) in AG media at pH 5 with increasing concentration of inositol was monitored at OD_{600}. The colors of the markers correspond to different concentrations of inositol as noted in the legend. (B) Maximum specific growth rate and lag time were quantified by fitting a single Gompertz to each growth curves from (A). The results are plotted according to increasing concentration of inositol. Arrows point to the concentration of inositol (100 mM) where the growth of the $\Delta hpnH$ strain is optimized. Error bars (standard deviation) are included, but some are smaller than the point markers.

Next, we tested the effects of divalent cations on growth (**Figure 5**). Because the

AG medium already has almost double the divalent cation concentration as the PSY

medium, we created a "low dication" pH 5 AG medium using the PSY concentrations of

divalent cations, specifically magnesium and calcium (45 μM Ca^{2+} and 400 μM Mg^{2+}). We then increased the calcium ion concentration in this low dication AG medium. The growth rate and lag time for WT and the $\Delta hpnH$ complement were agnostic to these changing conditions. However, the $\Delta hpnH$ strain grew at a slower rate and with a longer lag time in the low dication AG medium condition, indicating that 45 µM may be particularly stressful for the $\Delta hpnH$ strain. The growth rate recovered with 100 µM additional calcium and remained at the same growth rate. The lag time decreased as the calcium concentration was increased. When 500 µM additional calcium was added to the low divalent cation condition, the growth curve for $\Delta hpnH$ became almost indistinguishable from the pH 5 AG medium. These experiments were also completed with magnesium as the divalent cation for WT and the $\Delta hpnH$ strain with similar but less pronounced results (**Figure S4**).

Together, these results indicate that the lack of extended hopanoids can be compensated for by changing the physicochemical properties of the growth medium, specifically the osmolarity and divalent cation concentration. Unlike the pattern seen for osmolytes for $\Delta hpnH$, where an intermediate concentration minimized lag time and increased growth rate, increasing concentrations of divalent cations increasingly shrunk the lag time, yet did not appreciably affect growth rate.

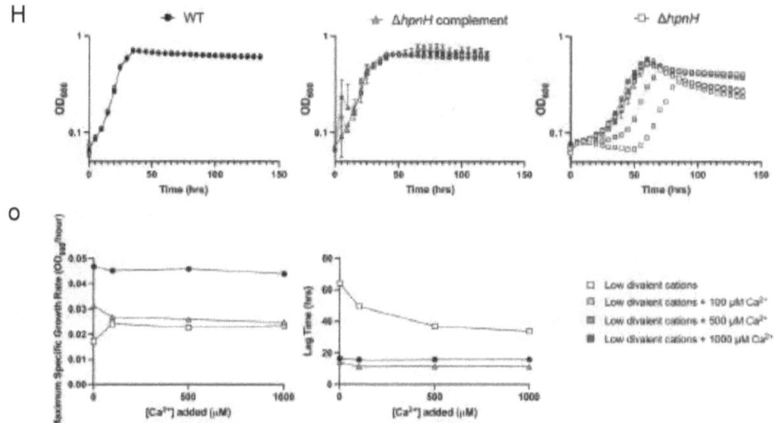

Figure 5. *B. diazoefficiens* Δ*hpnH* strain growth is sensitive to the concentration of calcium. (A) Growth of WT (black circles), Δ*hpnH* complement (grey triangles), and Δ*hpnH* (white squares) in AG media at pH 5 with different concentrations of divalent cations was monitored at OD_{600}. The colors of the markers correspond to different concentrations of Ca^{2+} ions as noted in the legend. (D) Maximum specific growth rate and lag time were quantified by fitting a single Gompertz curve to each growth curve from (C). The results are plotted according to increasing concentration of Ca^{2+} ions with the low dication condition included at y=0. All growth curves and quantifications represent the average of three biological replicates. Error bars (standard deviation) are included, but some are smaller than the point markers.

Extended hopanoids are required to regulate the membrane properties of B. diazoefficiens.

In order to explain the growth phenotypes of the Δ*hpnH* strain, we hypothesized that extended hopanoids play a role in buffering the outer membrane against physicochemical perturbations. Indeed, the unextended hopanoid, diplopterol, is known to modulate changes in lipid A packing *in vitro* that occurs in response to decreased pH (52). To probe the underlying mechanism behind the physicochemical compensation for loss of extended hopanoids, we used the lipophilic dyes Di-4-ANEPPGHQ (Di-4) and diphenyl hexatriene (DPH) (**Figure 5** and **S5**). The general polarization (GP) of Di-4 reports on lipid packing, with higher GPs indicative of increased packing. The fluorescence polarization of

DPH reports on membrane rigidity, which is coupled to lipid packing (53, 54). Because of their different sizes (665.55 MW for Di-4 vs. 232.33 MW for DPH) and polarities, Di-4 should preferentially label the outer leaflet of the outer membrane while DPH can permeate the outer membrane to label the inner membrane as well (55). However, given the interplay between membrane stability and permeability, these assumptions may not hold if membrane permeability is increased sufficiently to allow permeation of Di-4 across the outer membrane. Outer membranes contain saturated Lipid A and have been shown to have similar lipid packing to liquid ordered phase membranes in vitro (56) and inner membranes are comprised of more disordered phospholipids, which should have lower lipid packing similar to a liquid disordered phase. Thus, a large decrease in Di4 GP could either be interpreted as a decrease in outer membrane lipid packing, or to a large increase in outer membrane permeability allowing Di4 to label the inner membrane. Both results would indicate a large change in the mechanical properties of the outer membrane. Keeping these caveats in mind, Di-4 can be used to infer changes in outer membrane mechanical properties resulting from physicochemical perturbations.

To evaluate the role of extended hopanoids in outer membrane acclimation to pH and osmotic strength we compared Di-4 GP in cells grown at pH 6.6 and 5, and cells grown at pH 5 in the presence and absence of inositol. The Di-4 GP values for WT and the Δ*hpnH* complement strains were almost identical across the three conditions (pH 6.6 AG media, pH 5 AG media, and pH 5 AG media + 100 mM inositol), which indicates that the outer membrane of our complement strain is responding very similarly to WT. The Δ*hpnH* strain had lower GP values than the WT and the Δ*hpnH* complement when grown in pH 6.6 AG media, indicating that the Δ*hpnH* strain has a higher membrane fluidity, or possibly compromised membrane integrity allowing permeation of Di4 across the bilayer. This decreased membrane order for the Δ*hpnH* strain in pH 6.6 AG was confirmed by DPH polarization (**Figure S5**), suggesting that Di-4 GP is indicative of lower outer membrane lipid packing. Additionally, all three strains show an increase in lipid packing when grown in pH 5 AG medium, which is also expected due to the effect of pH on the density and order of LPS (52, 57). Interestingly, WT and the Δ*hpnH* complement show a small decrease in lipid packing in AG pH 5 medium with 100 mM inositol added, while the Δ*hpnH* strain lipid packing continues to increase. To better interpret these results, we determined the ΔGP values, comparing the change in GP compared to the standard pH 6.6 AG medium condition. Here, we observe that WT and the Δ*hpnH* complement strains underwent very small changes in lipid packing between conditions, but the Δ*hpnH* strain experienced much greater changes in lipid packing. Overall, these results indicate that extended hopanoids play an important role in buffering *B. diazoefficiens* membrane properties against physicochemical perturbations.

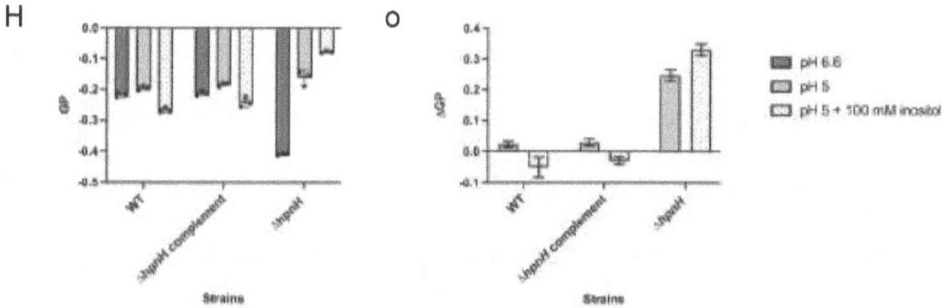

Figure 6. *B. diazoefficiens* Δ*hpnH* strain is deficient in its ability to regulate its membrane properties. (A) Di-4 general polarization (GP) measurements of WT, Δ*hpnH* complement, and Δ*hpnH* strains grown in AG media with pH 6.6 (dark grey), pH 5 (light grey), and pH 5 with 100 mM added inositol (grey dots). Individual measurements are shown as black circles. (B) ΔGP for each strain compared to the AG pH 6.6 condition. Error bars (standard deviation) are included.

Discussion

By making a serendipitous discovery of the conditional essentiality of the *shc* gene in *B. diazoefficiens*, we discovered that the physicochemical environment is extremely important for the growth of hopanoid-deficient *B. diazoefficiens* strains, including the extended hopanoid-deficient mutant, Δ*hpnH*. We confirmed that the Δ*hpnH* strain undergoes significant death in stationary phase, but that, in contrast to previous results (25), it can grow at pH 5 in a medium with higher osmolarity and divalent ion concentration. For the first time, we discovered environmentally relevant conditions that partially compensate for the *B. diazoefficiens* Δ*hpnH* mutant growth defect at low pH: intermediate osmolarity and elevated divalent cation concentrations. Finally, using a biophysical technique, we discovered that extended hopanoids are important for modulating lipid packing of the outer membrane.

When we first discovered that our conditional Δ*shc* mutant could grow on solid AG medium without cumate induced hopanoid production but not on PSY medium, it caused us to reexamine our previous results with Δ*hpnH* carried out in PSY (25). As previously shown, Δ*hpnH* has a stationary phase defect and increased lag time when grown in PSY. When additional stress was added to the PSY condition, such as increased temperature, lowered pH, or microooxia, the Δ*hpnH* strain was unable to grow at all. It is possible that due to its stationary phase defect under these conditions, the Δ*hpnH* inoculum in these experiments may have had fewer viable cells than WT, despite similar optical density measurements, contributing to the severity of these phenotypes. Indeed, the "control" growth of the Δ*hpnH* mutant is visually reduced compared to the WT in the stressor gradient plate assay performed in these studies, consistent with this hypothesis. With this in mind, our revised inoculation protocol enabled us to see that Δ*hpnH* grows very similarly to WT in AG medium and can even grow at pH 5. Surprisingly, Δ*hpnH* experiences significant death in stationary phase compared to WT in AG medium, despite the optical density measurements remaining constant. These results highlight the importance of not relying on optical density measurements to assess bacterial viability, as they can be decoupled, a known but often forgotten phenomenon. It is likely that the previous results in PSY were accentuated due to differences in inoculum viability, representing the combined effects of the stationary and exponential phase growth defects.

As we tried to understand why the two media affect the growth of our hopanoid-deficient mutants differently, we reflected on their composition. Three differences stood out to us: differences in pH, divalent cations, and osmolarity. Previous research on Δ*shc* mutants in other closely related bacteria have shown that hopanoids are important in both acidic and basic conditions. Specifically, a Δ*shc* mutant of *Rhodopseudomonas palustris* failed to grow as it made the medium more basic (58). We know that in both PSY and AG media, *B. diazoefficiens* increases the pH through amino acid metabolism of the complex media (i.e. yeast extract). However, the AG medium also contains a higher buffering capacity at a lower pH than PSY (6.6 vs 7), thus likely extending the time before the pH is increased substantially. Therefore, the conditional Δ*shc* mutant's lack of growth on PSY medium and growth on AG medium is perhaps not entirely surprising and illustrates how important medium composition may be when isolating and growing a mutant in hopanoid production or other membrane component.

The higher concentration of divalent cations, specifically magnesium and calcium, in the AG medium compared to PSY, was interesting because of the role these cations play in outer membrane cohesion. Specifically, magnesium and calcium intercalate within the LPS layer, shielding negatively charged residues, resulting in a more ordered and robust outer membrane in the face of physicochemical stressors (23, 49). Our two hopanoid-deficient mutants cannot make HoLA, a component of LPS that contributes to membrane ordering (27, 59). Recent work in *Bradyrhizobium* BTai1 has revealed that calcium ions increase membrane bilayer thickness, an indication of increased membrane order, in membrane vesicles containing LPS without a hopanoid attached (59); this phenomenon suggests a mechanism whereby calcium may be able to compensate for lack of hopanoids. Our work shows that the lag time of Δ*hpnH* in pH 5 AG medium can be reduced by increasing concentrations of divalent cations, supporting this hypothesis. Calcium has a greater effect than magnesium, likely reflecting the fact that calcium ions more strongly increase lipid bilayer rigidity through dehydration effects than magnesium ions (60). WT was agnostic to these changes in divalent ion concentration perhaps due to the presence of HoLA, unlike other Bradyrhizobia strains (61). Interestingly, while calcium is maintained at low concentrations (0.1 μM) in the cytosol of plant cells, calcium has been shown to localize to the root nodule (62). Calcium binding proteins were specifically found in the root nodules from *Medicago truncatula* (63). Indeed, sufficient calcium is needed for the bacteria to fix nitrogen in the root nodules (62, 64). This evidence illustrates the importance of calcium within the acidic root nodule (7), and helps rationalize why the Δ*hpnH* mutant grows reasonably well *in planta*: elevated calcium levels may compensate for the loss of extended hopanoids within the root nodule.

We were surprised when we found that the AG medium has a higher osmolarity than the PSY medium, since our previous *in vitro* studies in PSY medium indicated that hopanoids can be protective against hyperosmotic stress (25). Yet, as previously noted, these media are compositionally different in more than one way. It thus appeared possible that, at a lower pH, the relationship between osmolarity and hopanoids might be more nuanced. We hypothesized that hypoosmolarity might also be stressful for hopanoid-deficient mutants that have less robust membranes, and our findings bore this out. A potential mechanism that explains this observation follows: when first introduced to hypoosmotic conditions, water tries to move into the higher osmolarity cell, likely causing at least a transient increase in membrane fluidity as the cell stretches to accommodate the increased volume (65). The cell responds by opening mechanosensitive channels to eject solutes and lower the cytosolic osmolarity while synthesizing osmoregulated periplasmic glucans, thus osmotically buffering the cytoplasm (66). In a cell with a less robust membrane and perhaps increased permeability due to the absence of hopanoids, the increased fluidity during initial water influx may kill some cells, while the periplasmic glucans may be more easily lost to the medium, losing their ability to osmotically buffer the cytoplasm. At low pH, these effects would be magnified as the influx of water would also bring an influx of protons, adding an additional stress. When we added inositol to the pH 5 AG medium, the *ΔhpnH* strain grew better, decreasing the lag time and increasing growth rate up to 100 mM added inositol. Comparatively, WT grew more poorly upon even the smallest addition of inositol (25 mM). This result illustrates that the low osmolarity of the medium is particularly stressful to the *ΔhpnH* strain. Interestingly, inositol makes up a large proportion of the compounds found in the symbiosome space (67). Indeed, the symbiosome space contains approximately 180 mM of low molecular weight compounds (67), notably similar to the maximally restorative osmolarity in our experiments (100 mM inositol pH 5 AG medium). It is thus possible that the root nodule microenvironment allows the $\Delta hpnH$ strain to survive and fix nitrogen, despite its obvious growth defects at low pH.

Interestingly, while both osmolarity and divalent cation concentrations affect the growth of the $\Delta hpnH$ strain at low pH, the effects are different. Specifically, divalent cation concentration had the greatest effect on lag time while the added osmolytes affected both lag time and growth rate. These differences suggest that different mechanisms underpin the mutant's response, despite both having the potential to rigidify the outer membrane. We hypothesize that these differences may arise due to the inositol primarily addressing the root cause of the stress, hypoosmolartiy, while the increase in divalent cations protects against the effects of hypoosmolarity.

The sensitivity of hopanoid-deficient strains to specific external conditions suggested that extended hopanoids might be particularly important in helping cells respond to environmental changes. In contrast to diplopterol, a shorter hopanoid that contains a hydrophilic group, which has been shown to rigidify the membrane while keeping lipids from entering a gel phase and retaining lateral lipid diffusivity (52, 68) extended hopanoids had only been shown to rigidify the membrane (25, 69, 70). Our biophysical experiments confirm that extended hopanoids are necessary for membrane rigidification, but also reveal that lack of extended hopanoids causes greater problems with membrane stability. $\Delta hpnH$ displays much greater variability in lipid packing between conditions than WT, as evidenced by the larger ΔGP values for the mutant. This result suggests that the WT can adjust its lipid packing to maintain a relatively constant membrane fluidity. In contrast, the $\Delta hpnH$ strain struggles to adjust its lipid packing in response to environmental changes. The lipid packing of the $\Delta hpnH$ strain is primarily affected by the external environment. In the case of the pH 5 AG medium with 100 mM inositol added, both increased osmolarity and inositol specifically are known to rigidify membranes (65, 71), explaining the increased GP values for the Δ*hpnH* strain. On the other hand, the WT can adjust its membrane to counteract environmentally-triggered membrane rigidification, thus leading to slightly lower GP values. Overall, these results indicate that extended hopanoids play an important role in *B. diazoefficiens* adjustment to the external environment.

In conclusion, the lack of hopanoids, and specifically, extended hopanoids—which are required for HoLA biosynthesis—makes *B. diazoefficiens* particularly sensitivity to environmental conditions in ways that are relevant to its lifecycle. That the lack of extended hopanoids can be partially compensated for by a moderately high osmotic level, helps to resolve the paradox of why the Δ*hpnH* mutant can be symbiotically successful if given sufficient time to develop within root nodules. Yet, its sensitivity to hypoosmotic conditions suggest that hopanoids may provide a fitness advantage to rhizobia in waterlogged soils, where osmolytes and divalent cations are diluted. Together, our findings emphasize the importance of considering the full ecophysiological picture when attempting to understand the selective benefits of a given molecular component on an organism. It has been said that the only constant in life is change, a point worth remembering when considering the effects of hopanoids on peripatetic soil organisms.

References

1. Doran JW, Zeiss MR. 2000. Soil health and sustainability: managing the biotic component of soil quality. Applied Soil Ecology 15:3–11.
2. Norris CE, Congreves KA. 2018. Alternative management practices improve soil health indices in intensive vegetable cropping systems: A review. Front Environ Sci 6.
3. Tilman D, Cassman KG, Matson PA, Naylor R, Polasky S. 2002. Agricultural sustainability and intensive production practices. Nature 418:671–677.
4. Foyer CH, Nguyen H, Lam H-M. 2019. Legumes-The art and science of environmentally sustainable agriculture. Plant Cell Environ 42:1–5.
5. Oldroyd GED, Murray JD, Poole PS, Downie JA. 2011. The rules of engagement in the legume-rhizobial symbiosis. Annu Rev Genet 45:119–144.
6. Gibson KE, Kobayashi H, Walker GC. 2008. Molecular determinants of a symbiotic chronic infection. Annu Rev Genet 42:413–441.
7. Pierre O, Engler G, Hopkins J, Brau F, Boncompagni E, Hérouart D. 2013. Peribacteroid space acidification: a marker of mature bacteroid functioning in Medicago truncatula nodules. Plant Cell Environ 36:2059–2070.
8. Hunt S. 1993. Gas Exchange of Legume Nodules and the Regulation of Nitrogenase Activity. Annu Rev Plant Physiol Plant Mol Biol 44:483–511.
9. Tookmanian EM, Belin BJ, Sáenz JP, Newman DK. 2021. The role of hopanoids in fortifying rhizobia against a changing climate. Environ Microbiol 23:2906–2918.
10. Reckling M, Hecker J-M, Bergkvist G, Watson CA, Zander P, Schläfke N, Stoddard FL, Eory V, Topp CFE, Maire J, Bachinger J. 2016. A cropping system assessment framework—Evaluating effects of introducing legumes into crop rotations. European Journal of Agronomy 76:186–197.
11. Bullock DG. 1992. Crop rotation. CRC Crit Rev Plant Sci 11:309–326.
12. Loureiro M de F, Kaschuk G, Alberton O, Hungria M. 2007. Soybean [Glycine max (L.) Merrill] rhizobial diversity in Brazilian oxisols under various soil, cropping, and inoculation managements. Biol Fertil Soils 43:665–674.
13. Zhang NN, Sun YM, Li L, Wang ET, Chen WX, Yuan HL. 2010. Effects of intercropping and Rhizobium inoculation on yield and rhizosphere bacterial community of faba bean (Vicia faba L.). Biol Fertil Soils 46:625–639.
14. Roughley RJ, Gemell LG, Thompson JA, Brockwell J. 1993. The number of Bradyrhizobium SP. (Lupinus) applied to seed and its effect on rhizosphere

colonization, nodulation and yield of lupin. Soil Biol Biochem 25:1453–1458.

15. Corich V, Giacomini A, Vendramin E, Vian P, Carlot M, Concheri G, Polone E, Casella S, Nuti MP, Squartini A. 2007. Long term evaluation of field-released genetically modified rhizobia. Environ Biosafety Res 6:167–181.

16. O'Callaghan M. 2016. Microbial inoculation of seed for improved crop performance: issues and opportunities. Appl Microbiol Biotechnol 100:5729–5746.

17. Karmakar K, Rana A, Rajwar A, Sahgal M, Johri BN. 2015. Legume-Rhizobia Symbiosis Under Stress, p. 241–258. *In* Arora, NK (ed.), Plant microbes symbiosis: applied facets. Springer India, New Delhi.

18. Ilangumaran G, Smith DL. 2017. Plant growth promoting rhizobacteria in amelioration of salinity stress: A systems biology perspective. Front Plant Sci 8:1768.

19. Zahran HH. 1999. Rhizobium-legume symbiosis and nitrogen fixation under severe conditions and in an arid climate. Microbiol Mol Biol Rev 63:968–89, table of contents.

20. Delgado-Baquerizo M, Oliverio AM, Brewer TE, Benavent-González A, Eldridge DJ, Bardgett RD, Maestre FT, Singh BK, Fierer N. 2018. A global atlas of the dominant bacteria found in soil. Science 359:320–325.

21. Choma A, Komaniecka I, Zebracki K. 2017. Structure, biosynthesis and function of unusual lipids A from nodule-inducing and N2-fixing bacteria. Biochim Biophys Acta Mol Cell Biol Lipids 1862:196–209.

22. Serrato RV. 2014. Lipopolysaccharides in diazotrophic bacteria. Front Cell Infect Microbiol 4:119.

23. Nikaido H. 2003. Molecular Basis of Bacterial Outer Membrane Permeability Revisited. Microbiol Mol Biol Rev 67:593–656.

24. Komaniecka I, Zamłyńska K, Zan R, Staszczak M, Pawelec J, Seta I, Choma A. 2016. Rhizobium strains differ considerably in outer membrane permeability and polymyxin B resistance. Acta Biochim Pol 63:517–525.

25. Kulkarni G, Busset N, Molinaro A, Gargani D, Chaintreuil C, Silipo A, Giraud E, Newman DK. 2015. Specific hopanoid classes differentially affect free-living and symbiotic states of Bradyrhizobium diazoefficiens. MBio 6:e01251-15.

26. Belin BJ, Busset N, Giraud E, Molinaro A, Silipo A, Newman DK. 2018. Hopanoid lipids: from membranes to plant-bacteria interactions. Nat Rev Microbiol 16:304–315.

27. Silipo A, Vitiello G, Gully D, Sturiale L, Chaintreuil C, Fardoux J, Gargani D, Lee H-I, Kulkarni G, Busset N, Marchetti R, Palmigiano A, Moll H, Engel R, Lanzetta R, Paduano L, Parrilli M, Chang W-S, Holst O, Newman DK, Garozzo D, D'Errico G, Giraud E, Molinaro A. 2014. Covalently linked hopanoid-lipid A improves outer-membrane resistance of a Bradyrhizobium symbiont of legumes. Nat Commun 5:5106.

28. Busset N, Di Lorenzo F, Palmigiano A, Sturiale L, Gressent F, Fardoux J, Gully D, Chaintreuil C, Molinaro A, Silipo A, Giraud E. 2017. The Very Long Chain Fatty Acid (C26:25OH) Linked to the Lipid A Is Important for the Fitness of the PhotosyntheticBradyrhizobiumStrain ORS278 and the Establishment of a Successful Symbiosis withAeschynomeneLegumes. Front Microbiol 8:1821.

29. Komaniecka I, Choma A, Mazur A, Duda KA, Lindner B, Schwudke D, Holst O.

2014. Occurrence of an unusual hopanoid-containing lipid A among lipopolysaccharides from Bradyrhizobium species. J Biol Chem 289:35644–35655.

30. Belin BJ, Tookmanian EM, de Anda J, Wong GCL, Newman DK. 2019. Extended Hopanoid Loss Reduces Bacterial Motility and Surface Attachment and Leads to Heterogeneity in Root Nodule Growth Kinetics in a Bradyrhizobium-Aeschynomene Symbiosis. Mol Plant Microbe Interact 32:1415–1428.

31. Miller JH. 1972. Experiments in Molecular Genetics. Cold Spring Harbor Laboratory Pr, Cold Spring Harbor, N.Y.].

32. Mesa S, Hauser F, Friberg M, Malaguti E, Fischer H-M, Hennecke H. 2008. Comprehensive assessment of the regulons controlled by the FixLJ-FixK2-FixK1 cascade in Bradyrhizobium japonicum. J Bacteriol 190:6568–6579.

33. Regensburger B, Hennecke H. 1983. RNA polymerase from Rhizobium japonicum. Arch Microbiol 135:103–109.

34. Sadowsky MJ, Tully RE, Cregan PB, Keyser HH. 1987. Genetic Diversity in Bradyrhizobium japonicum Serogroup 123 and Its Relation to Genotype-Specific Nodulation of Soybean. Appl Environ Microbiol 53:2624–2630.

35. Cole MA, Elkan GH. 1973. Transmissible resistance to penicillin G, neomycin, and chloramphenicol in Rhizobium japonicum. Antimicrob Agents Chemother 4:248–253.

36. Kaczmarczyk A, Vorholt JA, Francez-Charlot A. 2013. Cumate-inducible gene expression system for sphingomonads and other Alphaproteobacteria. Appl Environ Microbiol 79:6795–6802.

37. Eaton RW. 1997. p-Cymene catabolic pathway in Pseudomonas putida F1: cloning and characterization of DNA encoding conversion of p-cymene to p-cumate. J Bacteriol 179:3171–3180.

38. Eaton RW. 1996. p-Cumate catabolic pathway in Pseudomonas putida Fl: cloning and characterization of DNA carrying the cmt operon. J Bacteriol 178:1351–1362.

39. Ledermann R. 2017. Role of general stress response in trehalose biosynthesis for functional rhizobia-legume symbiosis. Doctoral dissertation, ETH Zurich.

40. Gibson DG, Young L, Chuang R-Y, Venter JC, Hutchison CA, Smith HO. 2009. Enzymatic assembly of DNA molecules up to several hundred kilobases. Nat Methods 6:343–345.

41. Beck C, Marty R, Kläusli S, Hennecke H, Göttfert M. 1997. Dissection of the transcription machinery for housekeeping genes of Bradyrhizobium japonicum. J Bacteriol 179:364–369.

42. Masloboeva N, Reutimann L, Stiefel P, Follador R, Leimer N, Hennecke H, Mesa S, Fischer H-M. 2012. Reactive oxygen species-inducible ECF σ factors of Bradyrhizobium japonicum. PLoS One 7:e43421.

43. Hahn M, Meyer L, Studer D, Regensburger B, Hennecke H. 1984. Insertion and deletion mutations within the nif region of Rhizobium japonicum. Plant Mol Biol 3:159–168.

44. Ledermann R, Bartsch I, Remus-Emsermann MN, Vorholt JA, Fischer H-M. 2015. Stable Fluorescent and Enzymatic Tagging of Bradyrhizobium diazoefficiens to Analyze Host-Plant Infection and Colonization. Mol Plant Microbe Interact 28:959–967.

45. Tjørve KMC, Tjørve E. 2017. The use of Gompertz models in growth analyses, and

new Gompertz-model approach: An addition to the Unified-Richards family. PLoS One 12:e0178691.

46. Zwietering MH, Jongenburger I, Rombouts FM, van 't Riet K. 1990. Modeling of the bacterial growth curve. Appl Environ Microbiol 56:1875–1881.

47. Wu C-H, Bialecka-Fornal M, Newman DK. 2015. Methylation at the C-2 position of hopanoids increases rigidity in native bacterial membranes. Elife 4.

48. Lin ZF, Liu N, Lin GZ, Peng CL. 2011. Factors altering the membrane fluidity of spinach thylakoid as determined by fluorescence polarization. Acta Physiol Plant 33:1019–1024.

49. Clifton LA, Skoda MWA, Le Brun AP, Ciesielski F, Kuzmenko I, Holt SA, Lakey JH. 2015. Effect of divalent cation removal on the structure of gram-negative bacterial outer membrane models. Langmuir 31:404–412.

50. Schäfer A, Tauch A, Jäger W, Kalinowski J, Thierbach G, Pühler A. 1994. Small mobilizable multi-purpose cloning vectors derived from the Escherichia coli plasmids pK18 and pK19: selection of defined deletions in the chromosome of Corynebacterium glutamicum. Gene 145:69–73.

51. Bolan NS, Hedley MJ, White RE. 1991. Processes of soil acidification during nitrogen cycling with emphasis on legume based pastures. Plant Soil 134:53–63.

52. Sáenz JP, Sezgin E, Schwille P, Simons K. 2012. Functional convergence of hopanoids and sterols in membrane ordering. Proc Natl Acad Sci USA 109:14236–14240.

53. Ma Y, Benda A, Kwiatek J, Owen DM, Gaus K. 2018. Time-Resolved Laurdan Fluorescence Reveals Insights into Membrane Viscosity and Hydration Levels. Biophys J 115:1498–1508.

54. Steinkühler J, Sezgin E, Urbančič I, Eggeling C, Dimova R. 2019. Mechanical properties of plasma membrane vesicles correlate with lipid order, viscosity and cell density. Commun Biol 2:337.

55. Zgurskaya HI, López CA, Gnanakaran S. 2015. Permeability Barrier of Gram-Negative Cell Envelopes and Approaches To Bypass It. ACS Infect Dis 1:512–522.

56. Sáenz JP, Grosser D, Bradley AS, Lagny TJ, Lavrynenko O, Broda M, Simons K. 2015. Hopanoids as functional analogues of cholesterol in bacterial membranes. Proc Natl Acad Sci USA 112:11971–11976.

57. Brandenburg K, Seydel U. 1990. Investigation into the fluidity of lipopolysaccharide and free lipid A membrane systems by Fourier-transform infrared spectroscopy and differential scanning calorimetry. Eur J Biochem 191:229–236.

58. Welander PV, Hunter RC, Zhang L, Sessions AL, Summons RE, Newman DK. 2009. Hopanoids play a role in membrane integrity and pH homeostasis in Rhodopseudomonas palustris TIE-1. J Bacteriol 191:6145–6156.

59. Vitiello G, Oliva R, Petraccone L, Vecchio PD, Heenan RK, Molinaro A, Silipo A, D'Errico G, Paduano L. 2021. Covalently bonded hopanoid-Lipid A from Bradyrhizobium: The role of unusual molecular structure and calcium ions in regulating the lipid bilayers organization. J Colloid Interface Sci 594:891–901.

60. Papahadjopoulos D, Portis A, Pangborn W. 1978. Calcium-induced lipid phase transitions and membrane fusion. Ann N Y Acad Sci 308:50–66.

61. Macció D, Fabra A, Castro S. 2002. Acidity and calcium interaction affect the

growth of *Bradyrhizobium* sp. and the attachment to peanut roots. Soil Biol Biochem 34:201–208.

62. Izmailov SF. 2003. Calcium-Based Interactions of Symbiotic Partners in Legumes: Role of Peribacteroid Membrane. Russian Journal of Plant Physiology.

63. Liu J, Miller SS, Graham M, Bucciarelli B, Catalano CM, Sherrier DJ, Samac DA, Ivashuta S, Fedorova M, Matsumoto P, Gantt JS, Vance CP. 2006. Recruitment of novel calcium-binding proteins for root nodule symbiosis in Medicago truncatula. Plant Physiol 141:167–177.

64. Andreev IM, Andreeva IN, Dubrovo PN, Krylova VV, Kozharinova GM, Izmailov SF. 2001. Calcium Status of Yellow Lupin Symbiosomes as a Potential Regulator of Their Nitrogenase Activity: The Role of the Peribacteroid Membrane. Russian Journal of Plant Physiology.

65. Los DA, Murata N. 2004. Membrane fluidity and its roles in the perception of environmental signals. Biochim Biophys Acta 1666:142–157.

66. Miller KJ, Wood JM. 1996. Osmoadaptation by rhizosphere bacteria. Annu Rev Microbiol 50:101–136.

67. Tejima K, Arima Y, Yokoyama T, Sekimoto H. 2003. Composition of amino acids, organic acids, and sugars in the peribacteroid space of soybean root nodules. Soil Sci Plant Nutr 49:239–247.

68. Mangiarotti A, Genovese DM, Naumann CA, Monti MR, Wilke N. 2019. Hopanoids, like sterols, modulate dynamics, compaction, phase segregation and permeability of membranes. Biochim Biophys Acta Biomembr 1861:183060.

69. Kannenberg E, Blume A, McElhaney RN, Poralla K. 1983. Monolayer and calorimetric studies of phosphatidylcholines containing branched-chain fatty acids and of their interactions with cholesterol and with a bacterial hopanoid in model membranes. Biochimica et Biophysica Acta (BBA) - Biomembranes 733:111–116.

70. Chen Z, Sato Y, Nakazawa I, Suzuki Y. 1995. Interactions between bacteriohopane-32,33,34,35-tetrol and liposomal membranes composed of dipalmitoylphosphatidylcholine. Biol Pharm Bull 18:477–480.

71. Crowe LM, Mouradian R, Crowe JH, Jackson SA, Womersley C. 1984. Effects of carbohydrates on membrane stability at low water activities. Biochim Biophys Acta 769:141–150.

72. Casadaban MJ, Cohen SN. 1980. Analysis of gene control signals by DNA fusion and cloning in Escherichia coli. J Mol Biol 138:179–207.

73. Simon R, Priefer U, Pühler A. 1983. A Broad Host Range Mobilization System for In Vivo Genetic Engineering: Transposon Mutagenesis in Gram Negative Bacteria. Nat Biotechnol 1:784–791.

74. Ledermann R, Strebel S, Kampik C, Fischer H-M. 2016. Versatile Vectors for Efficient Mutagenesis of Bradyrhizobium diazoefficiens and Other Alphaproteobacteria. Appl Environ Microbiol 82:2791–2799.

Supplemental Material

Table S1. Strains, plasmids, and primers used in this study[a]

Strain, plasmid, or primer	Genotype, description, and/or construction	Source or Reference
Strains		
DH10	*Escherichia coli;* F⁻ *endA1 recA1 galE15 galK16 nupG rpsL ΔlacX74* Φ80*lacZΔM15 araD139 Δ(ara,leu)7697 mcrA Δ(mrr-hsdRMS-mcrBC)* λ⁻; DKN89	(72)
DH5α	*E. coli; supE44 ΔlacU169 (φ80 lacZΔM15) hsdR17 recA1 gyrA96 thi-1 relA2*	BRL, Gaithersburg, USA
S17-1	*E. coli; thi pro hdsR hdsM⁺ recA;* chromosomal insertion of RP4-2 (Tc::Mu Km::Tn7); DKN1	(73)
GM2163	*Escherichia coli;* St^r, Cm^r; F- araC14 leuB6 fhuA31 lacY1 tsx-78 glnV44(AS) galK2(Oc) galT22 λ– mcrA dcm-6 hisG4(Oc) rfbD1 rpsL136(StrR) dam-13::Tn9(CamR) xylA5 mtl-1 thi-1 mcrB1 hsdR2; DKN307	New England Biolabs
DKN1391	*B. diazoefficiens* 110*spc*4, Sp^r, WT	(33)
DKN1529	*B. diazoefficiens* 110*spc*4, Sp^r, Δ*hpnH*; deletion of blr3006 in DKN1391	(25)
LacZYA-Q1	*B. diazoefficiens* Tc^r 110*spc*4 Sp^r, Tc^r, pRJPcu1-lacZYA at *scoI* locus; chromosomal integration of cumate inducible *lacZ* operon	This study
DKN2510	*B. diazoefficiens* 110*spc*4, Sp^r, Tc^r, Δ*hpnH* P$_{aphII}$-*hpnH* at *scoI* locus; complementation of blr3006 in DKN1529	(30)
DKN1784	*B. diazoefficiens* 110*spc*4, Sp^r, Δ*shc* P$_{cu}$-*shc* at *scoI* locus; deletion of blr3004 in DKN1391 followed by complementation with cumate inducible promoter	This study
Plasmids		
pGK259	*shc* deletion vector; HindIII/PstI-digested blr3004 (*shc*) upstream and downstream fusion PCR product was ligated to HindIII/PstI-digested pK18*mobsacB* Km^r mobilizable pUC18 derivative, *mob*, *sacB*, (DKN1492)	(25)
pQH2	Tc^r pQH derivative with pBBR *oriV*	(36)
pRJPaph-lacZYA	Tc^r P$_{aphII}$-*lacZYA* for integration downstream of *scoI*	(74)
pQH2-Prrn-mut2	Tc^r (pQH2) *Bd*-P$_{rrn-mut2}$ between *cuO*	This study
pGK302	*shc* cumate conditional complement vector; Cloned *B. diazoefficiens shc* coding region into pRJPcu-lacZYA using SpeI/PstI sites	This study

pRJPaph-sYFP2	Constitutive sYFP2 expression vector; Tc[r] (pRJPaph-gfp_a1) Paph-*sYFP2* for integration downstream of the *scoI* locus.	(44)
Primers		
Prrn-mut2_f	GTACGTTGACAGCCCGGAAGGTGGGTGCTATAACCCC	
Prrn-mut2_r	GGGGTTATAGCACCCACCTTCCGGGCTGTCAAC	
Shccodfor-SpeI	TAT ATA TAA CTA GTA TGG ATT CCG TGA ACG CG	
Shccodrev-PstI	TAT ATA TAC TGC AGT CAC ATT CCG ACC CCT ACC	

[a]Km, Kanamycin; Sp, Spectinomycin; Tc, Tetracycline; Str, Streptomycin; Cam, chloramphenicol; [r] denotes resistance; [*]denotes genes which were codon optimized for GC-rich organisms (36).

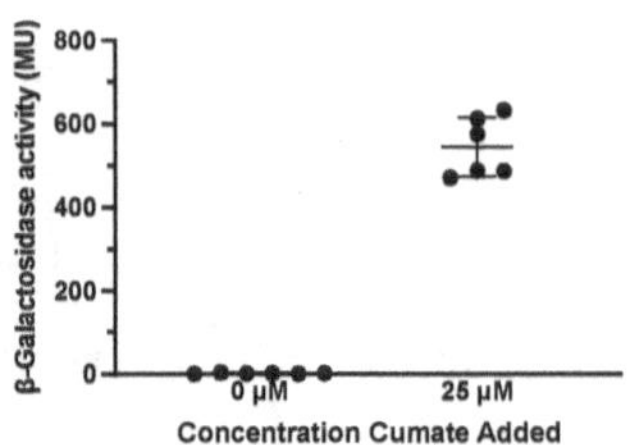

Figure S1. Cumate-inducible system functions as an on/off switch. For the strain LacZYA-Q1, when no cumate inducer is added, there is no β-Galactosidase activity. When 25 µM cumate inducer is added, β-Galactosidase activity is observed.

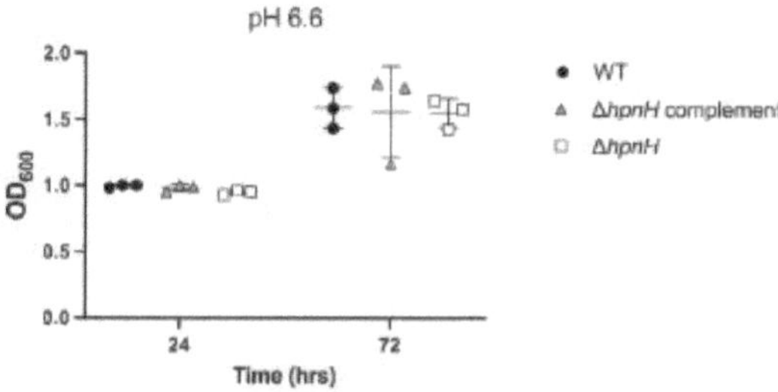

Figure S2. Differences in CFUs/mL not observed by OD_{600}. OD_{600} was measured for WT (circles), Δ*hpnH* complement (triangles), and Δ*hpnH* (squares) strains grown in AG media at pH 6.6 during exponential phase (24 hrs) and stationary phase (72 hrs). Error bars (standard deviation) are included, but some are obscured by the point markers.

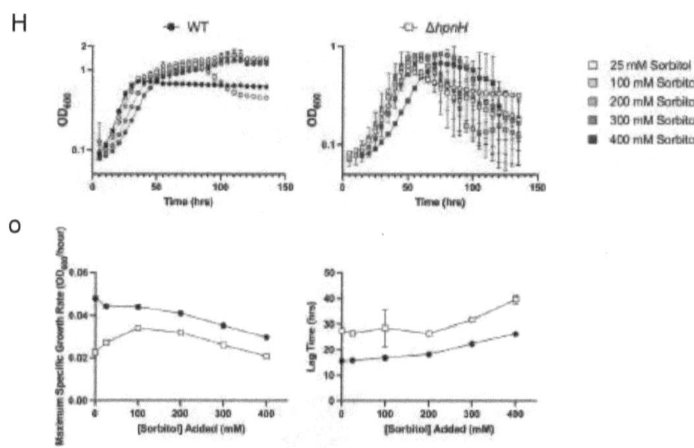

Figure S3. *B. diazoefficiens* Δ*hpnH* strain growth is sensitive to the concentration of sorbitol. (A) Growth of WT (circles), Δ*hpnH* complement (triangles), and Δ*hpnH* (squares) in AG media at pH 5 with increasing concentration of sorbitol was monitored at OD_{600}. The colors of the markers correspond to different concentrations of sorbitol as noted in the legend. (B) Maximum specific growth rate and lag were quantified by fitting a single Gompertz to each growth curve from (A). The results are plotted according to increasing concentration of sorbitol.

H

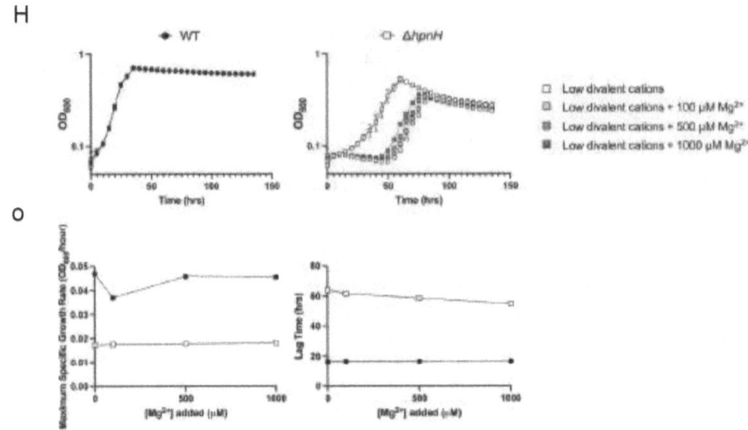

Figure S4. *B. diazoefficiens* Δ*hpnH* strain growth is sensitive to the concentration of magnesium ions. (A) Growth of WT (circles), Δ*hpnH* complement (triangles), and Δ*hpnH* (squares) in AG media at pH 5 with different concentrations of divalent cations was monitored at OD_{600}. The colors of the markers correspond to different concentrations of Mg^{2+} ions as noted in the legend. (B) μ and lag were quantified by fitting a single Gompertz curve to each growth curve from (A). The results are plotted according to increasing concentration of Mg^{2+} ions with the low dication condition included at y=0. All growth curves and quantifications represent the average of three biological replicates. Error bars (standard deviation) are included, but some are smaller than the point markers.

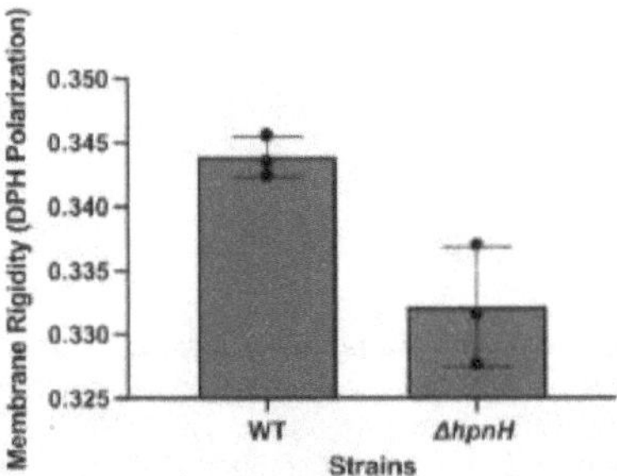

Figure S5. *B. diazoefficiens* Δ*hpnH* strain membrane is less ordered than WT when grown in pH 6.6 AG media. Whole-cell membrane fluidity measurements by fluorescence polarization of DPH. Error bars represent the standard deviations from three biological replicates.

Chapter 5

CONCLUSIONS

In this thesis, I have brought hopanoid research into the context of climate change. In the past, some hopanoid research has focused on understanding their role as biomarkers in the rock record (1). However, more recently, as the genetic ability to synthesize 2-methyl hopanoids was correlated to organisms known to interact with plants, the potential role of hopanoids in agriculture began to be explored (2). My work continues this investigation, examining the full lifecycle of rhizobia within the soil and the plant with an eye to the climate change we are currently experiencing.

In my very first group meeting presentation, I was asked what medium I used to grow my bacteria. I had not even thought to pay attention to what was in the PSY medium I had used because I did not understand the importance of the medium components for bacteria. However, over time, **medium matters** became the tagline of my thesis and opened my eyes to a different way of viewing my research: from the perspective of the bacteria. The medium, after all, is the controlled environment that a bacterium experiences in the lab, and the membrane is the first line of defense for bacteria from the environment. If hopanoid-deficient mutants were struggling in the controlled medium in the lab, how would they fair outside of the lab, in an environment that might be very different or change over time?

With this in mind, I became interested in the microenvironments that rhizobia would experience in their lifecycle, which I detailed in Chapter 2. I recognized that the symbiosome space within the root nodule is not the only stressful environment that rhizobia would encounter. The soil and rhizosphere can also be very stressful in part because the environment can fluctuate. Climate change will only make these conditions more intensely variable–especially due to expected changes in regional precipitation. I think understanding the soil and symbiosome environment at the scale of bacteria is an extremely important goal for future work. Without understanding the environment that is experienced in nature, our lab experiments are useless. Indeed, the paradox, introduced in Chapter 3, that the extended hopanoid-deficient mutant could develop a moderately successful symbiosis despite its defects in culture, emphasizes this point. The medium used in the culture experiments and the culturing method exacerbated growth defects that did not match the results *in planta*. As scientists look to engineer or select for rhizobia that improve the legume-rhizobia symbiosis, they need to make sure that their conditions are closely simulating the expected environments and consistently check phenotypes in soil/plant studies. Additionally, scientists should be selecting for climate change resistance, which is only possible if we understand the effects of climate change in the soil and the root nodule. The root nodule environment under conditions associated with climate change is extremely understudied and deserves attention.

The membrane intersects directly with the different environmental stresses rhizobia may encounter, including variable pH, hypo and hyperosmolarity, microoxia, and nutrient stress, because membrane physical properties affect membrane permeability, stability, and protein function. For this reason, mutants in membrane biosynthesis exhibit pleiotropic phenotypes that make a mechanistic understanding difficult. The way many phenotypes explained by the vague "membrane defect" without further examination felt like a limitation of studying membranes using bacteria to me. At the same time, *in vitro* lipid studies were also dissatisfying, since they ignored the complexity of the cell (i.e. protein interactions) and often rely on model lipids. In Chapter 4, by focusing on divalent cation concentration and osmolarity, I was able to tease apart some nuances of the role of hopanoids. Most importantly, I learned that extended hopanoids bestow robustness to different physichochemical conditions, a discovery that seems to encompass the many pleiotropic phenotypes exhibited. I believe that the robustness of the *B. diazoefficiens* membrane to physichochemical conditions deserves further investigation. Specifically, while diplopterol has been shown to maintain lateral lipid diffusivity and increase membrane rigidity, this property has not been tested for extended hopanoids such as bacteriohopanetetrol except for in modeling work (3–5). The results in Chapter 4 seem to hint that extended hopanoids may offer this property, but the absence of hopanoid attached lipid A (HoLA) in the extended hopanoid-deficient mutant complicates matters. I am interested in understanding what extent of these phenotypes are due to lack of HoLA vs. lack of extended hopanoids. *In vitro* experiments could help tease this apart, especially if using native lipid preparations. On the other hand, finding the enzyme responsible for attaching an extended hopanoid to lipid A would allow for further investigation of the role of HoLA in rhizobia through mutant construction and phylogenetic analysis—a priority for future research.

With my undergraduate background in protein work, I have also been very interested in how these changes in membrane properties affect membrane proteins. I began my work on hopanoids searching for hopanoid-binding proteins, detailed in the Appendix, and saw these effects in my motility studies, detailed in Chapter 3. In the Appendix, I have also outlined a potential path forward to find hopanoid-binding proteins, an important endeavor to understand how hopanoids specifically may impact protein function. However, more broadly, I am also interested in the effects of membrane composition on protein function. While membrane compositions and membrane proteins do seem to have coevolved to optimize protein function, there are no defined principles describing this interplay in the membrane (6). I would be interested in seeing a directed evolution experiment to try and understand the first principles of membrane protein and lipid interactions. Overall, I have been convinced by my research that lipids are understudied across fields and deserve more attention.

Finally, I think an underutilized tool to answer many biological questions is comparison across different bacterial species and strains. There has been research on hopanoids in many different species, which has offered generalized trends and information about the role of hopanoids. However, the experiments are usually carried out in different labs using different media and test different membrane stresses. To make matters more complicated, negative results may not be included. While none of these complaints invalidate the research that has been done previously, I think there is room for improvement in the future. Experiments that directly compare strains can be invaluable. For example, one of my favorite studies is an analysis of polymyxin resistance (an indication of outer membrane stability) across species of rhizobia (7). While mutants unable to make certain lipid A modifications showed decreased polymyxin resistance among other pleiotropic membrane defects (8, 9), comparing strains that "naturally" do not make the same modifications allowed for a more subtle comparison of membrane properties. Even when working with a single species, the choices scientists make can aid or hinder comparison across species. By completing the same experiments under similar conditions as previous researchers, future researchers will be able to make better comparisons, synthesize this information, and introduce new hypotheses. I believe the goal of making comparisons across species is very important to the future of microbiological research, especially as *E. coli* is slowly dethroned as the default bacterium and genome sequencing becomes cheaper.

My thesis has taught me many things, but I hope overall it illustrates how important a robust membrane is for bacteria. Lipids and membranes are extremely understudied, despite their importance. Soil bacteria live in an environment that undergoes many changes that will only become more extreme with climate change, but I think the lessons I have learned in this thesis are important when considering any environment. I was often frustrated by working with the membrane because its complexity defies a tidy explanation. However, I have come to realize that one of the reasons this work is so frustrating is because the membrane is so unbelievably important to the bacterial cell. For that reason, it has been worth the frustration to move this project forward through this thesis.

References
1. Newman DK, Neubauer C, Ricci JN, Wu C-H, Pearson A. 2016. Cellular and Molecular Biological Approaches to Interpreting Ancient Biomarkers. Annu Rev Earth Planet Sci 44:493–522.
2. Ricci JN, Michel AJ, Newman DK. 2015. Phylogenetic analysis of HpnP reveals the origin of 2-methylhopanoid production in Alphaproteobacteria. Geobiology 13:267–277.
3. Sáenz JP, Sezgin E, Schwille P, Simons K. 2012. Functional convergence of hopanoids and sterols in membrane ordering. Proc Natl Acad Sci USA 109:14236–14240.
4. Mangiarotti A, Genovese DM, Naumann CA, Monti MR, Wilke N. 2019. Hopanoids, like sterols, modulate dynamics, compaction, phase segregation and permeability of membranes. Biochim Biophys Acta Biomembr 1861:183060.
5. Caron B, Mark AE, Poger D. 2014. Some like it hot: the effect of sterols and hopanoids on lipid ordering at high temperature. J Phys Chem Lett 5:3953–3957.
6. Amin DN, Hazelbauer GL. 2012. Influence of membrane lipid composition on a transmembrane bacterial chemoreceptor. J Biol Chem 287:41697–41705.
7. Komaniecka I, Zamłyńska K, Zan R, Staszczak M, Pawelec J, Seta I, Choma A. 2016. Rhizobium strains differ considerably in outer membrane permeability and polymyxin B resistance. Acta Biochim Pol 63:517–525.
8. Ingram BO, Sohlenkamp C, Geiger O, Raetz CRH. 2010. Altered lipid A structures and polymyxin hypersensitivity of Rhizobium etli mutants lacking the LpxE and LpxF phosphatases. Biochim Biophys Acta 1801:593–604.
9. Brown DB, Huang Y-C, Kannenberg EL, Sherrier DJ, Carlson RW. 2011. An acpXL mutant of Rhizobium leguminosarum bv. phaseoli lacks 27-hydroxyoctacosanoic acid in its lipid A and is developmentally delayed during symbiotic infection of the determinate nodulating host plant Phaseolus vulgaris. J Bacteriol 193:4766–4778.

Appendix

TOWARD IDENTIFYING HOPANOID-BINDING PROTEINS

Versions of this work first appeared as a candidacy research summary (May 2017) and independent research proposal (June 2021).

Introduction

Over the years, hopanoids have been constantly compared to cholesterol due to their similar structure (**Figure 1**). Indeed, the function of hopanoids in bacteria has echoed a well-known role for cholesterol in eukaryotes–to maintain membrane robustness under stressful conditions (1). As a planar molecule, cholesterol interacts well with the n-C_{16-18} chains of eukaryotic phospholipids, leading to cohesive packing based on cooperative van der Waals interactions (2). Cholesterol can help the membrane condense and become more rigid and stable, but cholesterol also increases lateral membrane fluidity which helps maintain membrane protein function (3). *In vitro* work has confirmed that many hopanoids, including bacteriohopanetetrol (BHT) can condense model membranes (4, 5), while diplopterol can also maintain the lateral diffusivity of membrane lipids, much like cholesterol (6, 7).

However, the role of cholesterol within the cell extends beyond maintaining membrane properties. Cholesterol is known to bind proteins and modulate protein function (8). Lipid-protein interactions in the membrane are ubiquitous since proteins make up 50-75% (w/w) of the membrane and these proteins must be sufficiently hydrophobic to reside in the membrane (9). Cholesterol has a bumpy (β) and a smooth (α) side, which leads to specific interactions with proteins. As different interactions have been characterized, different cholesterol-binding domains (CRAC, CARC, and TILT domains) have been identified (8). These advances allow for prediction of cholesterol-interacting proteins. In contrast, very little work has explored whether hopanoids bind proteins specifically like cholesterol, making this an area ripe for investigation.

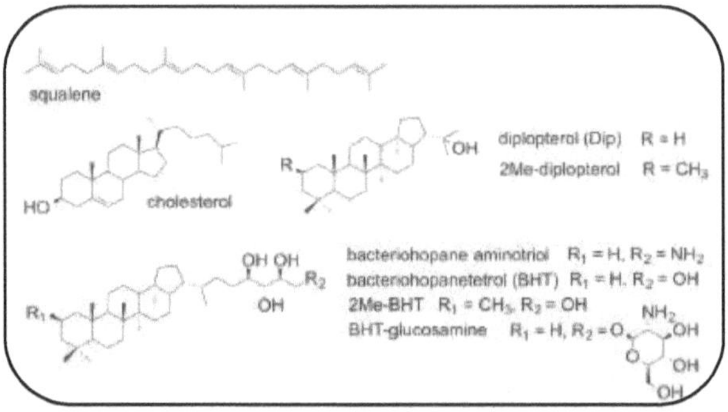

Figure 1. Various hopanoid structures and cholesterol. Adapted from (49).

Cholesterol and its derivatives are known to interact with some proteins as signaling molecules. For example, in higher organisms they play a role in Hedgehog signaling and development (10). More fundamentally, they are known to bind sterol regulatory element containing proteins, to inhibit transcriptional activators of lipid synthesis (11). However, unlike cholesterol, whose degradation is relatively well understood, a hopanoid degradation pathway has not been identified in any organism (12, 13). If there are no mechanisms to regulate the concentration of hopanoids within the cell, hopanoids could be a poor signaling molecule. Additionally, cholesterol is the precursor for other important biomolecules like vitamin D and hormones, but hopanoids, while diverse in their structure, are the end of the biosynthetic pathway (14). These differences indicate that though hopanoids may act as signaling molecules, they may not act in the same manner or with the same purpose as cholesterol.

Cholesterol is also a key component of ordered membrane domains, also known as lipid rafts (15–17). The fluid mosaic model has framed the understanding of membranes for decades, leading to the discovery of long-range order within eukaryotic membranes. This heterogeneous distribution of membrane lipids and proteins, called membrane domains, are functionally important to organize proteins for sorting and trafficking, facilitating cell division, or signal transduction, indicating that the lipids involved may interact specifically with certain proteins for recruitment to these domains (8, 18, 19). Membrane domains were thought to be limited to eukaryotes because formation depended on the presence of cholesterol and the lingering of the untrue paradigm that eukaryotes are more advanced and complex organisms than bacteria or archaea. However, recent research on the membrane-associated sensor kinase KinC in *Bacillus subtilis* lead to the discovery of these domains in bacteria that contain polyisoprenoid lipids–precursors of hopanoids (20, 21). Hopanoids have since been implicated in membrane ordered domains in bacteria, mostly through *in vitro* work with diplopterol, which showed their ability to form liquid ordered domains (6). Cardiolipin, has also been implicated in bacterial and eukaryotic membrane domains through the use of cardiolipin specific dyes (19, 22). Importantly, cardiolipin is known to increase in the course of lipid remodeling in hopanoid-deficient cells, potentially confirming an overlap in function for hopanoids and cardiolipin in membrane domains (23).

It is tempting to think only of the similarities between cholesterol and hopanoids, but they have very different ring structures and modifications (24). In fact, these differences lead to different interactions with unsaturated phospholipids. Cholesterol interacts favorably with unsaturated phospholipids, while hopanoids actually have an unfavorable interaction (25). It is important to investigate the possible similarities with awareness of the ways these two molecules may work differently.

There are some proteins that have been hypothesized as candidate hopanoid-binding proteins. First, multi-drug efflux pumps have been implicated as interacting with hopanoids. Though many of the stress response defects in hopanoid-deficient mutants reported have been interpreted to illustrate hopanoid's role in maintaining membrane properties, sensitization to detergents is also a classic phenotype for defects in multidrug transport systems. Further testing in *Methylobacterium extorquens,* showed that energy-dependent multidrug transport is deficient in the Δ*shc* mutant (25). Additionally, motility has been shown to be defective in multiple hopanoid-deficient mutants (26, 27). Motility involves the synthesis and function of the flagellum, a large protein complex inserted through both the inner and outer membrane. While it is possible that these defects are related generally to membrane properties, these hopanoids may interact specifically with efflux pumps or flagella to modulate their cellular function.

Another potential hopanoid-binding protein is the tryptophan-rich sensory protein/translocator protein (TSPO) (28). Found in all three domains of life, the function of TSPO is unclear, but generally it has been shown to be involved in various stress responses. A specific cholesterol binding motif was identified that is well conserved for eukaryotes, with more variability in bacteria, perhaps to accommodate the different structure of hopanoids (29). Despite this well documented hypothesis, only cholesterol and tetrapyrroles have been rigorously tested for binding (30). TSPO represents a possible positive control worth testing as well as it could illustrate how some proteins may have evolved linearly from binding hopanoids to binding cholesterol.

The structural similarity between hopanoids and cholesterol has led to confirmation of many shared functions, but there is potential for these shared roles to be expanded. The following work begins to address potential protein-hopanoid interactions through an affinity proteomics assay with proposed improvements.

Materials and Methods

Bacterial strains and culture media

B.diazoefficiens 110spc4, Sp^r WT strain was provided as a gift from Hans Martin-Fischer (31). *B. diazoefficiens* was grown shaking at 250 rpm aerobically at 30°C in rich medium (peptone-salts-yeast extract medium with 0.1% arabinose (PSY) (31, 32)). For plates, media was solidified with 1.5% (w/v) agar.

Synthesizing hopanoid (+) resin

Lingbing Kong in Stuart Conway's group (Oxford University, UK) carried out this work. Literature procedures were followed to synthesize diploptene (DPT) from hopanone extracted from dammar resin (33). Diploptene was further functionalized to an aldehyde before it was covalently attached to Purolite amino C$_6$ acrylate resin (EC8405) (**Figure 2**).

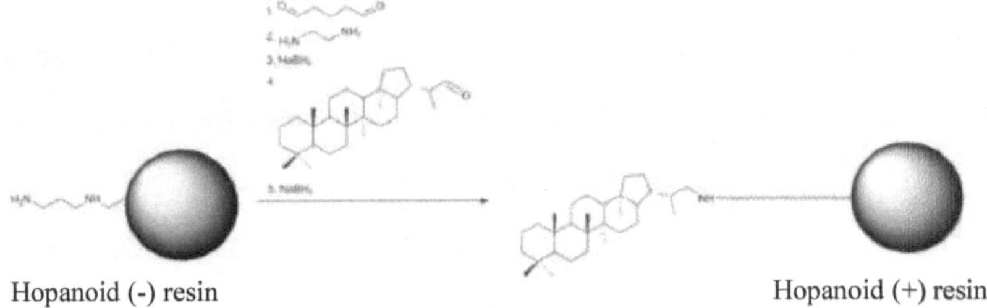

Hopanoid (-) resin   Hopanoid (+) resin

Figure 2. Synthesis of hopanoid (+) resin. Figure courtesy of Lingbing Kong.

Harvesting and lysing cells

Wildtype (WT) *B. diazoefficiens* was grown in PSY and harvested in early stationary phase (OD_{600} ~1). The cells were stored at -80°C until they were resuspended in lysis buffer (50 mM HEPES, 300 mM KCl, 2 mM EDTA, 0.02% DDM, 2 mM DDT, 0.1 mg/mL lysozyme, and one Roche Protease inhibitor tablet). After resuspension, 2 mM $MgCl_2$ and 8 µg/g pellet DNAseI was added. Lysis was carried out in 5-6 rounds through the Emulsiflex at 20,000 psi. The cells were spun down for 20 minutes at 10,000 g to remove cell debris. For some experiments, the remaining lysate was aliquoted half for cleared lysate (35 minutes at 16,000 g) and half for ultracentrifugation (1.5 hours at 90,000 g) to separate membrane and cytosol fraction. The membrane faction was resuspended in lysate buffer with 1% Triton X-100.

Binding capacity

Hopanoid (+) resin was incubated with cleared cell lysate for 5 minutes at room temperature before removing the supernatant. The amount of protein in the supernatant was quantified by a Biorad Bradford assay. This was repeated 15 times until saturation was observed followed by three 5-minute washes with wash buffer.

Affinity enrichment procedure

The resins were washed with acetonitrile and then equilibrated with wash buffer (50 mM HEPES, 300 mM KCl, 2 mM EDTA). Approximately 6 mg of both hopanoid (+) and hopanoid (-) resin with proteins from the respective fractions for one hour at room temperature. The different lysates were diluted to 1.6 mg/mL and incubated with 8-16 mg of lysate. The resin was spun down and the supernatant was removed followed by a single wash with wash buffer. The resin samples were then stored at -80°C before further processing.

Mass spectrometry workflow

The proteins were removed from the resins by denaturing the proteins by boiling and separation by SDS 4-12% Bis-Tris gel. In-gel trypsin digest was followed by desalting and quantification of peptides on HPLC. The method of dimethyl isotopic labeling was used to quantitatively compare the proteins enriched on the hopanoid (+) and hopanoid (-) resin in a single mass spectrometry run. The hopanoid (+) samples N-termini were chemically labeled with isotopically heavy or light methyl groups, while the hopanoid (-) samples N-termini were labeled with the opposite isotope pattern. The labeled samples were desalted by HPLC or C18 column and then mixed at a 1:1 ratio based on the HPLC quantification. At this point a PEG clean up method was attempted on some samples, which involved a C18 column with additional 0.2% formic acid in dichloromethane washes. All samples were run on the mass spectrometers at the Proteomic Exploration Lab at Caltech. The Orbitrap Fusion and Orbitrap Elite were used for different samples. The resulting mass spectra were analyzed with MaxQuant to identify proteins by their peptide fragments and assign isotopic ratios between the hopanoid (+) and hopanoid (-) enrichments.

Results and Discussion

Affinity probe synthesis and protocol refinement

To identify proteins that could interact with hopanoids, a hopanoid-functionalized resin (hopanoid (+) resin) was utilized to enrich for putative hopanoid binders. The hopanoid (-) resin was functionalized to create hopanoid (+) resin by our collaborator, Lingbing Kong. First, an approximate protein binding capacity for the hopanoid (+) resin was determined. The calculated binding capacity was ~50 μg protein/mg resin; however, as observed in **Figure 3**, the binding was inconsistent and took fifteen 50 μg incubations to reach saturation. To test the effects of washing, three washes with buffer were completed. The initial wash removed ~6% of proteins bound, indicating removal of non-specific binders or proteins in the small volume of liquid that remains associated with the resin. Subsequent washes removed 2-3% of proteins bound, most likely indicating removal of less specific or lower affinity binders. Based on these results, only one washing step was included in subsequent pulldowns and the resin was incubated with a protein excess of 100x the binding capacity (5 mg). General trouble shooting of the initial pulldown procedure was completed by Lingbing Kong and Cajetan Neubauer, a postdoc in the Newman lab.

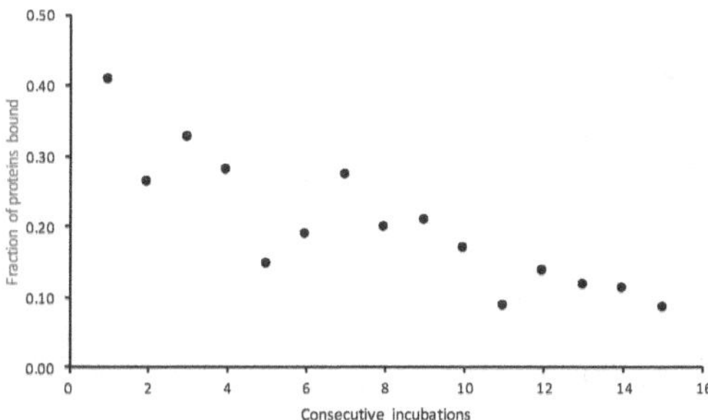

Figure 3. Hopanoid (+) resin binding affinity assay. 50 μg proteins from frozen *B. diazoefficiens* cleared lysate was applied at each incubation. Proteins removed in supernatant were quantified by a Bradford assay.

Hopanoid affinity enrichment

Wildtype *B. diazoeffiens* cells grown to late exponential phase were harvested and stored at -80°C before they were lysed and separated into different fractions. The pulldown experiment was completed three times with cleared lysate sample and one time with cytosolic and membrane fractions, generated by ultracentrifugation from one of the lysate samples. The general workflow is shown in **Figure 4**. Both hopanoid (+) and hopanoid (-) resins were incubated with proteins from the respective fractions for one hour at room temperature. The resin samples were then stored at -80°C before further standard processing for proteomics. The method of dimethyl isotopic labeling was used to quantitatively compare the proteins enriched on the hopanoid (+) and hopanoid (-) resin in a single mass spectrometry run. Proteins were identified by their peptide fragments and assigned isotopic ratios between the hopanoid (+) and hopanoid (-) enrichments.

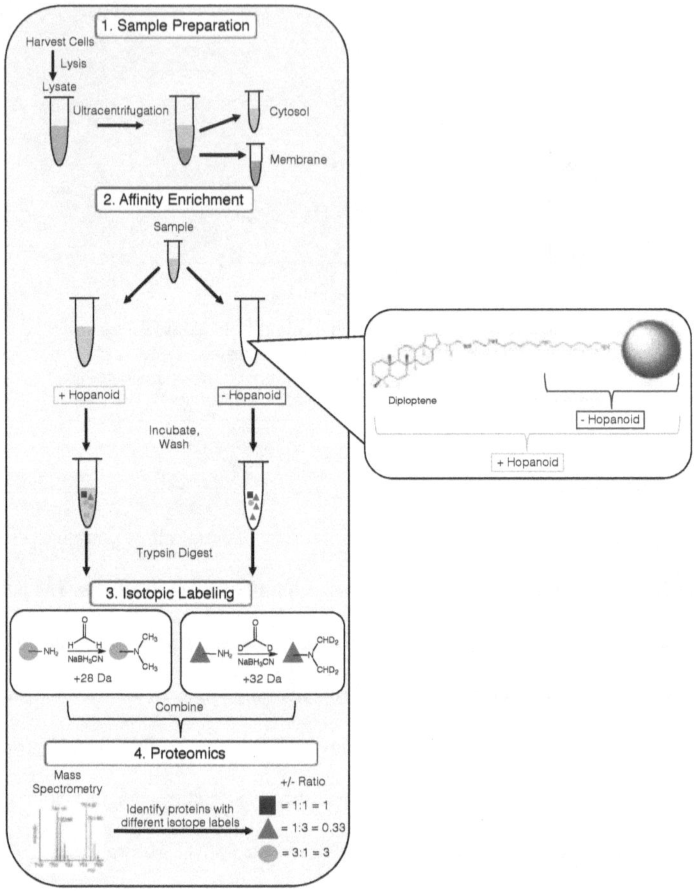

Figure 4. Workflow for hopanoid affinity enrichment of potential hopanoid-interacting proteins.

Analysis

Promisingly, in all experiments, more proteins were in the hopanoid (+) sample than the hopanoid (-) sample by gel staining. We were able to identify proteins enriched in the hopanoid (+) sample by large ratios comparing hopanoid (+) to hopanoid (-). Overall,

all samples showed an increase in the median ratio above 1. Since the two protein samples were mixed at a 1:1 ratio, any increase in the ratio indicates that these proteins are enriched on the hopanoid (+) bead and are likely interacting favorably with the hopanoids on the bead. Unfortunately, the distribution of ratios is a long-tailed distribution, so even large ratios were not statistically significant. Adding to this difficulty, due to slight variations in the protocols used, the results cannot be analyzed as strict biological replicates, although they are analyzed together (see below).

Two of our three lysate samples had polyethyleneglycol (PEG) contamination that suppresses ionization, lowering the number of peptides and proteins that could be identified and quantified. 413 proteins were identified and given ratios in the uncontaminated sample (replicate 1) compared to 71 and 40 replicate 2 and 3, respectively. When comparing these three samples, only 58 proteins were shared between replicate 1 and 2, 31 were shared between replicate 1 and replicate 3, and 19 were shared between replicate 2 and 3. When the shared proteins percentiles in each list were compared, the percentiles between

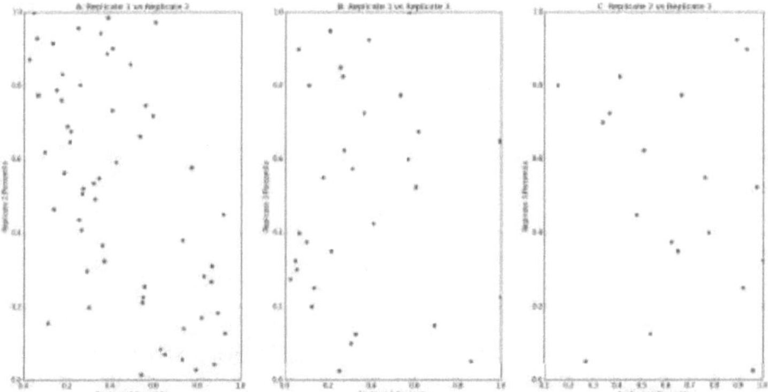

Figure 5. Analysis of lysate replicates. Percentiles were calculated based on empirical cumulative distribution functions of the entire sample's identified proteins. Percentiles were compared for all proteins that were present in both samples.

replicates are not correlated (**Figure 5**). These results indicate a lack of reproducibility, likely due to the PEG contamination which reduced the number of proteins quantified.

The sample that was fractionated into a cleared lysate, cytosol, and membrane fraction was not contaminated with PEG, and many proteins were identified and quantified in each fraction (413, 414, 396, respectively). The overlap in proteins identified between these samples was greater than between the three lysate samples with 212 proteins shared between all three samples. This indicates that either the PEG contamination or the difference between biological replicates is greater than the difference between the different fractions. There is broad but qualitative correlation of the percentiles of shared proteins between the three fractions (**Figure 6**). Based on this analysis, only the results from the fractionated sample were considered further. The results from the fractionated sample were treated as weighted lists with larger ratios indicative of a higher probability that these

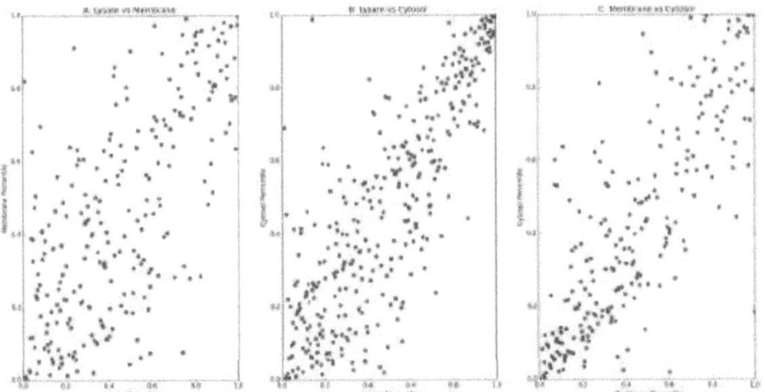

Figure 6. Analysis of fractionated sample. Percentiles were calculated based on empirical cumulative distribution functions of the entire fraction's identified proteins. Percentiles were compared for all proteins that were present in both fractions.

proteins are having a real interaction with the hopanoids on the bead. The top 10% of hits from each fraction are shown in a table in the supplemental material.

Initial attempts to carry out a hopanoid affinity enrichment identified some possible hopanoid interacting proteins, but overall, due to lack of reproducibility, indicated that the protocol needs to be majorly improved to discover hopanoid-binding proteins with any confidence. To that end, I have conceived improved strategies to find hopanoid-binding proteins which I have outlined below.

Proposed Future Work

Synthesize and validate bifunctional extended hopanoid probes

As the importance of understanding lipid-binding proteins has been recognized, there have been great strides in the development of chemical-biological approaches to identify these proteins. We have adapted the strategies previously developed for cholesterol and fatty acids for hopanoids (34–36). We have designed bifunctional probes containing a photoreactive moiety (diazirine) and a clickable handle (alkyne) (**Figure 8** and **Supplemental Material**). The diazirine covalently crosslinks with probe-interacting proteins upon UV light irradiation. The alkyne can be conjugated to azide-reporter tags by copper-catalyzed azide-alkyne cycloaddition (click) chemistry for detection, enrichment, and identification of probe-interacting proteins. The covalent crosslinking is a great improvement over relying on noncovalent interactions. We varied the placement of these moieties on the probes to account for different binding modes that might, for example, make the alkyne inaccessible for a click reaction. The probes have been designed to mimic the structure of BHT, a good target for identifying general hopanoid-binding proteins since hopanoids with a hydrophilic group like BHT are likely to adopt an upright orientation within the lipid bilayer, most relevant to protein-binding that has been seen for cholesterol (8, 24).

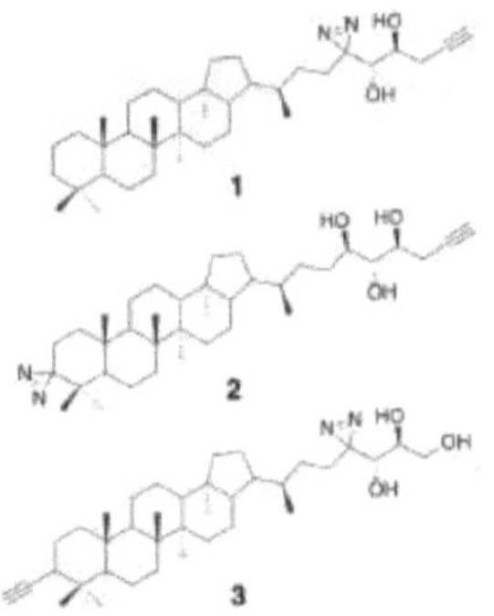

Figure 8. Bifunctional extended hopanoid probes. Three bifunctional hopanoid probes that mimic the structure of BHT are shown with varied positions of an alkyne and diazirine moiety. The synthetic routes are in the supplemental material.

The overall experimental setup is laid out in **Figure 9**. In brief, our synthesized probes will be added to live cells to bind target proteins. After UV irradiation, the probe reacts via photocrosslinking with the target protein. Cells are lysed and click chemistry is carried out on the samples to attach a fluorophore or affinity label for downstream analysis such as SDS-PAGE or quantitative proteomics. For mass spectrometry, we will conjugate our probe-crosslinked proteins to an azide0biotin tag by click chemistry and enrich these proteins using streptavidin chromatography. After trypsin digest, our samples will be labeled using tandem mass tags (TMT) for quantitative proteomics (37, 38). This quantitative proteomics approach involves differentially labeling peptides from multiple conditions so that they can be combined and compared in the same mass spectrum. For

SDS-PAGE, the probe will be conjugated to an azide-rhodamine dye for visualization of proteins.

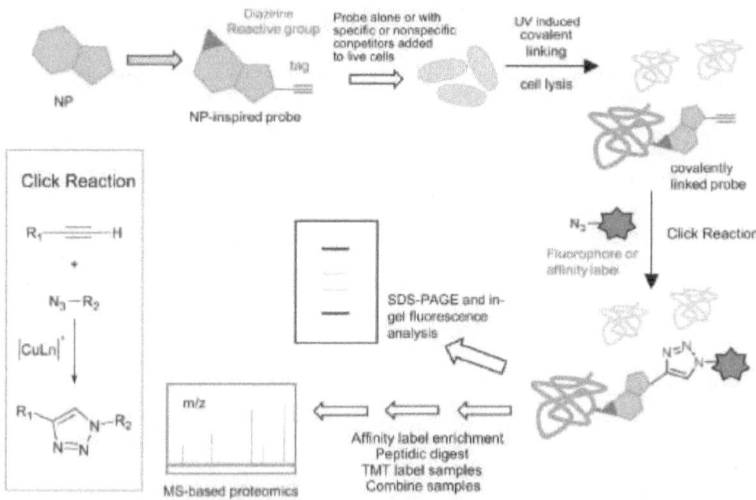

Figure 9. Experimental setup. A natural product (NP) inspires a probe with an alkyne tag (blue) and diazirine reactive group (red triangle). The probe is added to live cells and binds target proteins. Probe binding can be competed with either a specific competitor (the NP) or a nonspecific competitor (similar but different molecule). The probe then reacts via photocrosslinking with the target protein. Cells are lysed and click chemistry is carried out on the samples to attach a fluorophore or affinity label for downstream analysis such as SDS-PAGE or quantitative proteomics. Adapted from (38).

To validate our probe, we will check that the protein populations do not change significantly due to incubation with our probes by mass spectrometry. Next, we will check the efficiency of our photo crosslinking, alkyne accessibility, and wash steps by SDS-PAGE. We will compare whole cell samples incubated with our probe in complex with methyl-β-cyclodextrin with or without UV irradiation. Finally, we will test our probes for binding specificity. We will use a competition experiment where after incubation with our probe, we will add excess synthesized BHT. If our probes bind proteins specifically, then the excess BHT should directly compete for protein-binding, leading to fewer proteins visualized by SDS-PAGE. To further test if our probe binding is specific, we could look specifically at proteins that we expect to bind hopanoids (e.g. TSPO, HpnH, HpnG, and SHC). After expressing and purifying these proteins, we could compare binding of our probe and hopanoids (BHT and diploptene) using isothermal calorimetry.

From the results of these experiments, we could rationally alter the design of our probes if needed. One possible change could be adjusting the placement of the diazirine or alkyne. If we are unable to identify probe-binding proteins by SDS-PAGE when incubating whole cells, we could also try cell lysates with mild detergents such as n-dodecyl-D-maltoside (DDM).

Proteome mapping of putative hopanoid binding proteins

At this point, we will be poised to find putative hopanoid binding proteins in *Bradyrhizobium diazoefficiens* or other strains (**Figure 9**). We will compare samples incubated with our probe to samples that also incubated with a nonspecific competitor (cholesterol or a fatty acid) or a specific competitor (BHT). These comparisons will give us different groups of proteins that may be further investigated. By using TMT, we can quantitatively compare these three different conditions.

As a complementary or alternative approach, we can use a modified in cell thermal shift assay (CETSA) to identify potential hopanoid-binding proteins (**Figure 10**) (39). This assay is usually limited to cytosolic proteins, but there has been success using mild detergent to extend this approach to include membrane proteins (40). We expect many hopanoid-binding proteins to be membrane proteins, so this will be an important modification to our protocol. CETSA relies on the idea that binding small molecules often stabilizes protein structures, increasing their resistance to heat-induced aggregation. Variations of this approach have been used to determine lipid-protein binding and stabilization, but they have been more focused on analyzing specific proteins (41–43). For our approach, we would incubate cells with or without BHT and incubate at increasing temperatures. Using TMT, we could compare up to sixteen temperatures together and extract approximate melt curves of various proteins. By comparing the approximate melt curves +/- BHT, we could look for proteins whose melting temperatures increase significantly with addition of BHT. We would specifically look at the same proteins that we expect to bind hopanoids (e.g. HpnH, HpnG, and SHC) and validate individual protein hits from our results.

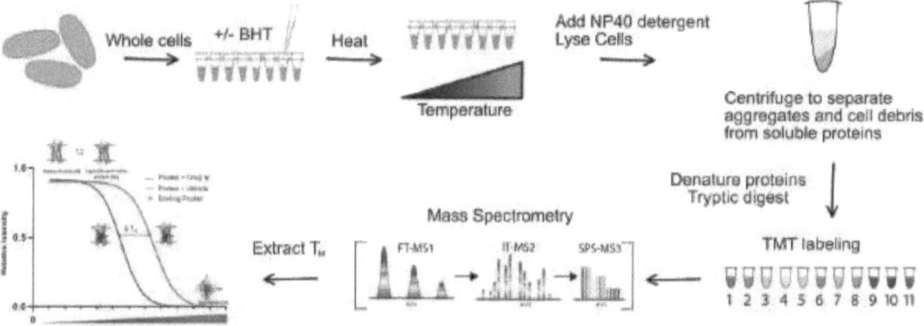

Figure 10. Unbiased cellular thermal shift assay using multiplexed quantitative mass spectrometry. Whole cells are aliquoted and incubated with or without BHT. Each aliquot is heated to a specific temperature before adding detergent and lysing the cells. Next the samples are centrifuged to isolate the soluble proteins. The soluble protein samples are denatured, digested, TMT labelled, and pooled before mass spectrometry. From the mass spectra, the melting temperature (T_M) of various proteins can be determined. Hopanoid binding will be determined by T_M increase when BHT is included. Adapted from (37, 39, 41).

Conclusion

By looking for potential hopanoid-binding proteins, we will discover more about the role hopanoids play in protein binding and potentially other aspects of barrier function maintenance. The area of hopanoid-binding proteins is ripe for investigation, but as I learned during this project, discovering these proteins is full of challenges. I hope this work helps anyone hoping to pursue this avenue in the future.

Supplemental Material

Table S1. Putative hopanoid-interacting proteins.

The top 10% of proteins based on their ratios identified in the lysate, cytosol, or membrane fractions. List is sorted by the largest ratios occurring in the largest number of samples.

Uniprot ID	Annotation	Lysate Ratio	Membrane Ratio	Cytosol Ratio
P53575	Electron transfer flavoprotein subunit beta	125.5	14.6	195.9
Q89XT2	Acetoacetyl CoA reductase	66.6	9.6	64.9
Q89IN8	ABC transporter substrate-binding protein	22.9	3.3	36.4
Q89IU7	Threonine--tRNA ligase	16.8	2.8	27.9
Q89LQ5	Two-component response regulator	20.2	3.2	22.0
Q89XI2	Bll0332 protein	21.0	4.5	20.0
Q89R53	N-carbamoyl-beta-alanine amidohydrolase	18.4	8.7	16.3
Q89GX2	3-hydroxybutyryl-CoA dehydrogenase	13.3	5.3	22.1
Q89J50	SAM-dependent methyl transferase	12.9	7.2	18.2

Q89JB0	short-chain dehydrogenases/reductases (SDR) family	10.0	8.5	17.8
Q89MV9	3-oxoacyl-(Acyl carrier protein) reductase	10.8	5.4	16.6
Q89QC3	Dehydrogenase	15.2	4.2	13.4
Q89C33	D-3-phosphoglycerate dehydrogenase	7.2	5.1	16.2
Q89QG9	Possible Acetyl-CoA synthetase (ADP-forming) alpha chain	9.0	1.8	16.2
Q89RD9	Hypothetical sugar kinase	6.0	11.8	7.9
Q89I92	Isocitrate dehydrogenase [NADP]	14.7	3.8	5.6
Q89WR6	Nitrogen regulatory protein PII	10.8	4.7	8.4
Q89N70	Inosine-5-monophosphate dehydrogenase	8.1	7.4	8.3
Q89HP5	S-adenosylmethionine synthase	9.0	5.1	9.2
Q89CK2	Two-component response regulator	7.0	6.5	9.6
Q89H21	Lon protease	12.3	3.9	6.2
Q89J83	30S ribosomal protein S10	7.4	7.3	6.2
Q89NC7	Dihydroxy-acid dehydratase	4.0	9.4	4.5
Q89F12	Transcriptional regulatory protein	24.2	2.3	NA
Q89EF9	NAD(P)+ transhydrogenase	13.1	10.0	NA
Q89VE9	Acetylornithine aminotransferase 1	18.3	3.3	NA
Q89RJ1	3-isopropylmalate dehydrogenase	13.7	2.6	NA
Q89KJ1	NADH ubiquinone oxidoreductase chain E	5.8	6.6	NA
Q89WH0	Oxidoreductase	6.8	5.0	NA
Q89X59	Malate dehydrogenase	2.0	5.1	NA
Q89EJ2	3-oxoadipate CoA-transferase subunit A	NA	4.5	17.2
Q89X71	ATP synthase subunit delta	NA	19.3	1.8
Q89VL6	Two-component response regulator	NA	10.0	6.2
Q89E05	AmiC protein	NA	5.2	8.2
P51130	Ubiquinol-cytochrome c reductase iron-sulfur subunit	NA	5.0	5.2
Q89QX0	hpnE - squalene-associated flavin adenine dinucleotide (FAD)-dependent desaturase	NA	5.1	3.1
Q89HN5	Formyltetrahydrofolate deformylase	35.2	NA	34.6
Q89IK9	10 kDa chaperonin	41.1	NA	16.7
Q89QU6	Putative oxidoreductase	34.0	NA	16.2
Q89UK8	Hypothetical protein - possible signal peptide	20.9	NA	16.9
Q89C45	4-hydroxy-tetrahydrodipicolinate synthase	20.9	NA	15.9
Q89SQ4	CheW protein	20.4	NA	12.8
Q89S80	Hypothetical protein - TPR repeat containing	19.9	NA	13.6
Q89N39	Putative activator of Hsp90 ATPase 1 family protein	20.4	NA	10.9
Q89V62	Two-component response regulator	16.2	NA	18.0

Q89UB8	UTP-glucose-1-phosphate uridylyltransferase	15.3	NA	19.3
Q89C57	Aminopeptidase	17.9	NA	11.3
Q89V88	Acyl-CoA thiolase	13.1	NA	15.3
Q89DC1	Inositol monophosphatase family protein	12.7	NA	13.2
Q89CC9	ABC transporter substrate-binding protein	14.3	NA	9.1
Q89KZ2	Similar to arginyl-tRNA synthetase	12.4	NA	9.7
Q89UU2	Peroxiredoxin	6.7	NA	19.9
Q89NC9	3-oxoacyl-(Acyl-carrier protein) reductase	11.1	NA	10.1
Q89FV0	D-alanine--D-alanine ligase	12.0	NA	5.6
Q89XV0	Peptidylprolyl isomerase	11.2	NA	6.9
Q89UV5	Dehydrogenase	8.1	NA	12.8
Q89DR2	Carbamoyl-phosphate synthase large chain	1.6	NA	23.9
Q89H30	Hypothetical protein	10.6	NA	5.7
Q89J68	Transcription termination/antitermination protein NusG	16.5	NA	NA
Q89DN5	Hypothetical glutathione S-transferase like protein	12.9	NA	NA
Q89G29	Hypothetical glutathione S-transferase like protein	12.9	NA	NA
Q89NT7	Two-component response regulator	11.3	NA	NA
Q89X27	3-isopropylmalate dehydratase small subunit	NA	16.0	NA
Q89KH0	czcC outer membrane protein, cobalt-zinc-cadmium efflux system	NA	13.2	NA
Q89H66	Phasin protein	NA	11.5	NA
Q89K46	Large-conductance mechanosensitive channel	NA	10.6	NA
Q89VM2	Amino acid ABC transporter substrate-binding protein, PAAT family	NA	10.0	NA
Q89MS3	Nucleoside diphosphate kinase	NA	8.6	NA
Q89UA6	Hypothetical transmembrane protein	NA	8.4	NA
Q89MV6	Endolytic murein transglycosylase (YceG family)	NA	8.3	NA
Q89H40	ABC transporter ATP-binding protein	NA	6.9	NA
Q89W89	Probable sugar kinase	NA	6.1	NA
Q89KQ1	Outer membrane protein assembly factor BamA	NA	6.1	NA
Q89WD1	Probable protein kinase UbiB	NA	5.8	NA
Q89UA7	Flagellar motor protein	NA	5.5	NA
Q89R98	ABC transporter ATP-binding protein	NA	5.3	NA
Q89N04	ABC transporter ATP-binding protein	NA	5.1	NA
Q89NJ1	Cytosine deaminase	NA	5.0	NA

Q89DN8	D-3-phosphoglycerate dehydrogenase	NA	NA	19.7
Q89CJ3	3-hydroxyacyl-CoA dehydrogenase type II	NA	NA	19.0
Q89HL2	SAM-dependent methyl transferase	NA	NA	18.5
Q89GX9	Glutathione-dependent formaldehyde-activating enzyme	NA	NA	15.7
Q89WK1	2,3-bisphosphoglycerate-dependent phosphoglycerate mutase	NA	NA	14.7
Q89VX0	Beta-ketoadipyl CoA thiolase	NA	NA	13.8
Q89I88	Aminomethyltransferase	NA	NA	13.7
H7C6H5	Hypothetical heme utilization protein	NA	NA	13.2
Q89LK8	Dihydroxy-acid dehydratase 1	NA	NA	13.0
Q89IH5	Uncharacterized MobA-related protein	NA	NA	12.8
O69161	Ribonuclease 3	NA	NA	11.9

Synthetic Schemes for Extended Hopanoid Bifunctional Probes

Scheme 1.

Scheme 2.

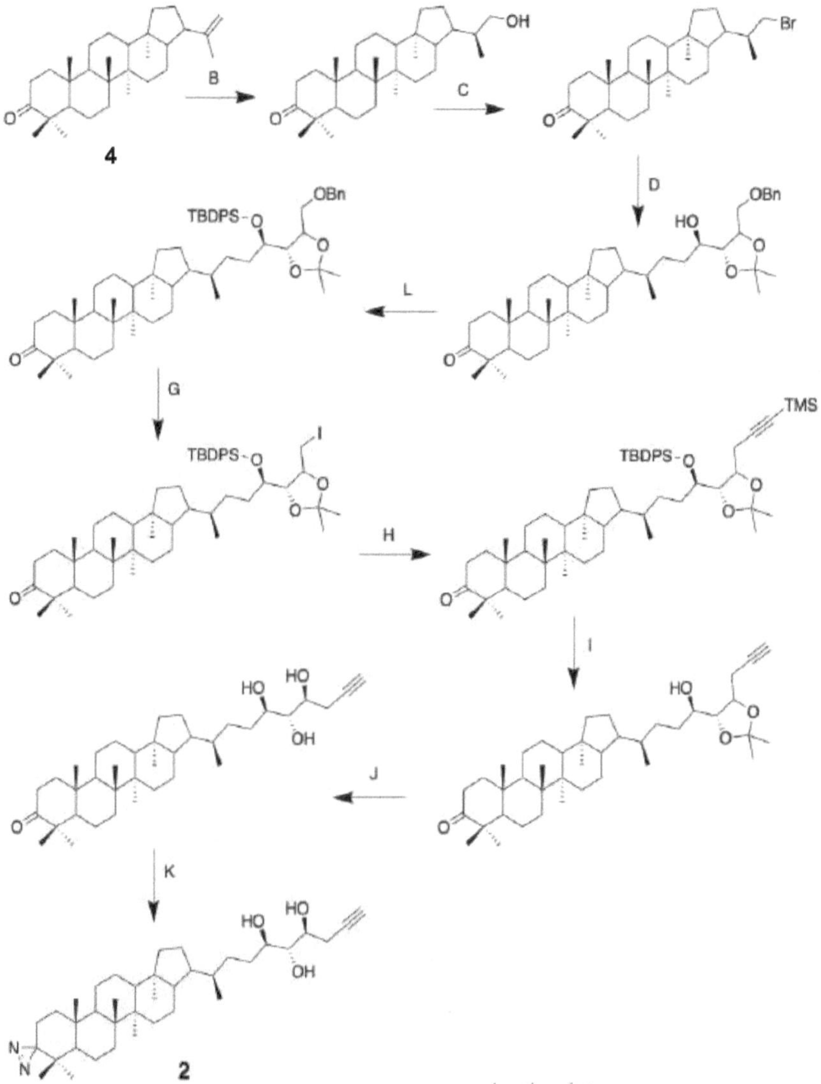

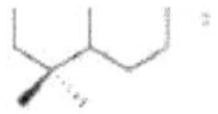

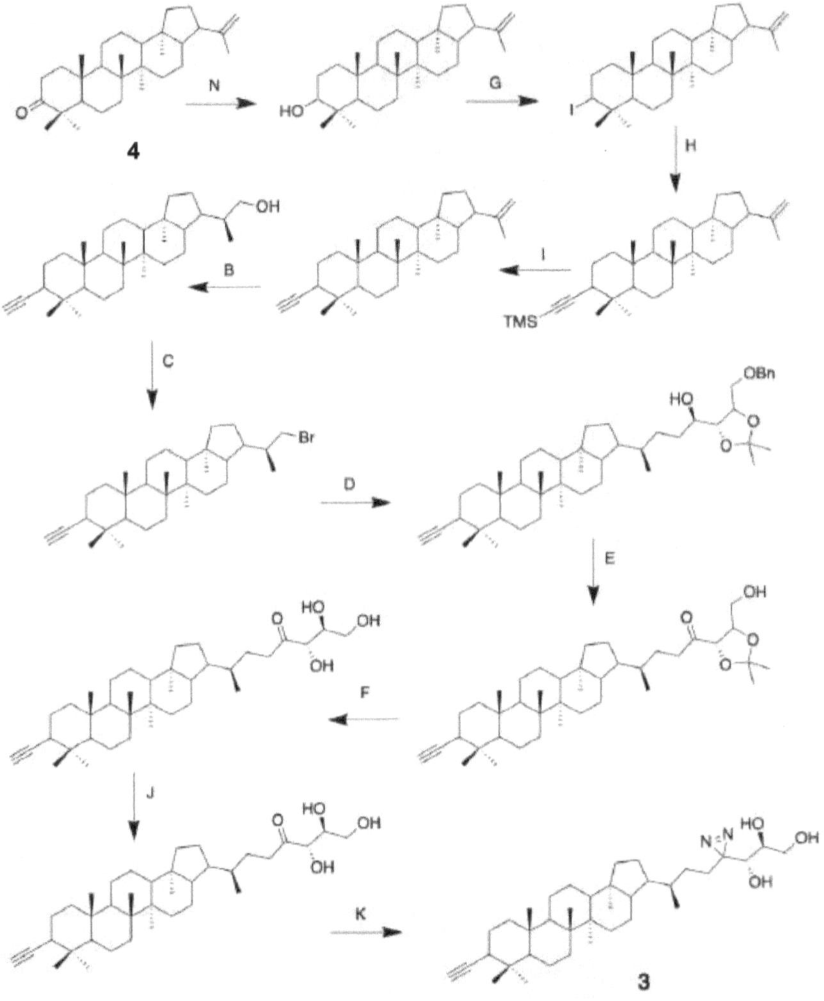

Scheme 3.

Reaction Conditions:

Synthesis of **4**: Hydroxyhopanone will be extracted from Dammar resin and then transformed to **4** using PdCl$_2$ in anhydrous toluene with 3 Å molecular sieves at 70°C (33).

A. NaBH$_4$, anhydrous toluene/MeOH with 3 Å molecular sieves, reflux (33)
B. 9-BBN, THF, RT; NaOH/H$_2$O$_2$ (44)
C. PPh$_3$, Br$_2$ (45)
D. (Thienyl)CuCNLi, lithium naphthalenide, (45)
E. PCC, DCM (46)
F. H$_2$, Pd(OH)$_2$ (45)
G. PPh$_3$, imidazole, I$_2$, DCM (36)
H. TMS———H , [(π-allyl)PdCl]$_2$, CuI, Cs$_2$CO$_3$, DMF/Et$_2$O, (36)
I. NaOH, THF/H$_2$O (36)
J. Camphorsulfonic acid (45)
K. (i) NH$_3$ in MeOH, 0°C (ii) NH$_2$OSO$_3$H, MeOH, 0°C → RT (iii) I$_2$, Et$_3$N, MeOH, RT (36)
L. DMAP, TBDPSCl, pyridine, 0°C → RT (47)
M. TBAF in acetic acid, THF (48)
N. NaBH$_4$, MeOH/Et$_2$O (33)

References

1. Belin BJ, Busset N, Giraud E, Molinaro A, Silipo A, Newman DK. 2018. Hopanoid lipids: from membranes to plant-bacteria interactions. Nat Rev Microbiol 16:304–315.
2. Demel RA, De Kruyff B. 1976. The function of sterols in membranes. Biochimica et Biophysica Acta (BBA) - Reviews on Biomembranes 457:109–132.
3. Nagimo A, Sato Y, Suzuki Y. 1991. Electron spin resonance studies of phosphatidylcholine interacted with cholesterol and with a hopanoid in liposomal membrane. Chem Pharm Bull 39:3071–3074.
4. Kannenberg E, Blume A, McElhaney RN, Poralla K. 1983. Monolayer and calorimetric studies of phosphatidylcholines containing branched-chain fatty acids and of their interactions with cholesterol and with a bacterial hopanoid in model membranes. Biochimica et Biophysica Acta (BBA) - Biomembranes 733:111–116.
5. Chen Z, Sato Y, Nakazawa I, Suzuki Y. 1995. Interactions between bacteriohopane-32,33,34,35-tetrol and liposomal membranes composed of dipalmitoylphosphatidylcholine. Biol Pharm Bull 18:477–480.
6. Sáenz JP, Sezgin E, Schwille P, Simons K. 2012. Functional convergence of hopanoids and sterols in membrane ordering. Proc Natl Acad Sci USA 109:14236–14240.
7. Mangiarotti A, Genovese DM, Naumann CA, Monti MR, Wilke N. 2019. Hopanoids, like sterols, modulate dynamics, compaction, phase segregation and permeability of membranes. Biochim Biophys Acta Biomembr 1861:183060.
8. Fantini J, Barrantes FJ. 2013. How cholesterol interacts with membrane proteins: an exploration of cholesterol-binding sites including CRAC, CARC, and tilted domains. Front Physiol 4:31.
9. Newman DK, Neubauer C, Ricci JN, Wu C-H, Pearson A. 2016. Cellular and Molecular Biological Approaches to Interpreting Ancient Biomarkers. Annu Rev Earth Planet Sci 44:493–522.
10. Beachy PA, Cooper MK, Young KE, von Kessler DP, Park WJ, Hall TM, Leahy DJ, Porter JA. 1997. Multiple roles of cholesterol in hedgehog protein biogenesis and signaling. Cold Spring Harb Symp Quant Biol 62:191–204.
11. Horton JD. 2002. Sterol regulatory element-binding proteins: transcriptional activators of lipid synthesis. Biochem Soc Trans 30:1091–1095.
12. Drzyzga O, Fernández de las Heras L, Morales V, Navarro Llorens JM, Perera J. 2011. Cholesterol degradation by Gordonia cholesterolivorans. Appl Environ Microbiol 77:4802–4810.
13. VanderVen BC, Fahey RJ, Lee W, Liu Y, Abramovitch RB, Memmott C, Crowe AM, Eltis LD, Perola E, Deininger DD, Wang T, Locher CP, Russell DG. 2015. Novel inhibitors of cholesterol degradation in Mycobacterium tuberculosis reveal how the bacterium's metabolism is constrained by the intracellular environment. PLoS Pathog 11:e1004679.
14. Payne AH, Hales DB. 2004. Overview of steroidogenic enzymes in the pathway from cholesterol to active steroid hormones. Endocr Rev 25:947–970.
15. Kim J, London E. 2015. Using Sterol Substitution to Probe the Role of Membrane

Domains in Membrane Functions. Lipids 50:721–734.

16. Ahmed SN, Brown DA, London E. 1997. On the origin of sphingolipid/cholesterol-rich detergent-insoluble cell membranes: physiological concentrations of cholesterol and sphingolipid induce formation of a detergent-insoluble, liquid-ordered lipid phase in model membranes. Biochemistry 36:10944–10953.

17. Brown DA, London E. 1998. Structure and origin of ordered lipid domains in biological membranes. J Membr Biol 164:103–114.

18. Mileykovskaya E, Dowhan W. 2005. Role of membrane lipids in bacterial division-site selection. Curr Opin Microbiol 8:135–142.

19. Bramkamp M, Lopez D. 2015. Exploring the existence of lipid rafts in bacteria. Microbiol Mol Biol Rev 79:81–100.

20. López D, Fischbach MA, Chu F, Losick R, Kolter R. 2009. Structurally diverse natural products that cause potassium leakage trigger multicellularity in Bacillus subtilis. Proc Natl Acad Sci USA 106:280–285.

21. López D, Kolter R. 2010. Functional microdomains in bacterial membranes. Genes Dev 24:1893–1902.

22. Mileykovskaya E, Dowhan W. 2009. Cardiolipin membrane domains in prokaryotes and eukaryotes. Biochim Biophys Acta 1788:2084–2091.

23. Neubauer C, Dalleska NF, Cowley ES, Shikuma NJ, Wu CH, Sessions AL, Newman DK. 2015. Lipid remodeling in Rhodopseudomonas palustris TIE-1 upon loss of hopanoids and hopanoid methylation. Geobiology 13:443–453.

24. Poger D, Mark AE. 2013. The relative effect of sterols and hopanoids on lipid bilayers: when comparable is not identical. J Phys Chem B 117:16129–16140.

25. Sáenz JP, Grosser D, Bradley AS, Lagny TJ, Lavrynenko O, Broda M, Simons K. 2015. Hopanoids as functional analogues of cholesterol in bacterial membranes. Proc Natl Acad Sci USA 112:11971–11976.

26. Belin BJ, Tookmanian EM, de Anda J, Wong GCL, Newman DK. 2019. Extended Hopanoid Loss Reduces Bacterial Motility and Surface Attachment and Leads to Heterogeneity in Root Nodule Growth Kinetics in a Bradyrhizobium-Aeschynomene Symbiosis. Mol Plant Microbe Interact 32:1415–1428.

27. Schmerk CL, Bernards MA, Valvano MA. 2011. Hopanoid production is required for low-pH tolerance, antimicrobial resistance, and motility in Burkholderia cenocepacia. J Bacteriol 193:6712–6723.

28. Li F, Liu J, Liu N, Kuhn LA, Garavito RM, Ferguson-Miller S. 2016. Translocator Protein 18 kDa (TSPO): An Old Protein with New Functions? Biochemistry 55:2821–2831.

29. Li F, Liu J, Valls L, Hiser C, Ferguson-Miller S. 2015. Identification of a key cholesterol binding enhancement motif in translocator protein 18 kDa. Biochemistry 54:1441–1443.

30. Busch AWU, WareJoncas Z, Montgomery BL. 2017. Tryptophan-Rich Sensory Protein/Translocator Protein (TSPO) from Cyanobacterium Fremyella diplosiphon Binds a Broad Range of Functionally Relevant Tetrapyrroles. Biochemistry 56:73–84.

31. Regensburger B, Hennecke H. 1983. RNA polymerase from Rhizobium japonicum. Arch Microbiol 135:103–109.

32. Mesa S, Hauser F, Friberg M, Malaguti E, Fischer H-M, Hennecke H. 2008.

Comprehensive assessment of the regulons controlled by the FixLJ-FixK2-FixK1 cascade in Bradyrhizobium japonicum. J Bacteriol 190:6568–6579.

33. Wu CH, Kong L, Bialecka-Fornal M, Park S, Thompson AL, Kulkarni G, Conway SJ, Newman DK. 2015. Quantitative hopanoid analysis enables robust pattern detection and comparison between laboratories. Geobiology 13:391–407.

34. Lum KM, Sato Y, Beyer BA, Plaisted WC, Anglin JL, Lairson LL, Cravatt BF. 2017. Mapping Protein Targets of Bioactive Small Molecules Using Lipid-Based Chemical Proteomics. ACS Chem Biol 12:2671–2681.

35. Hulce JJ, Cognetta AB, Niphakis MJ, Tully SE, Cravatt BF. 2013. Proteome-wide mapping of cholesterol-interacting proteins in mammalian cells. Nat Methods 10:259–264.

36. Peng T, Hang HC. 2015. Bifunctional fatty acid chemical reporter for analyzing S-palmitoylated membrane protein-protein interactions in mammalian cells. J Am Chem Soc 137:556–559.

37. O'Connell JD, Paulo JA, O'Brien JJ, Gygi SP. 2018. Proteome-Wide Evaluation of Two Common Protein Quantification Methods. J Proteome Res 17:1934–1942.

38. Wright MH, Sieber SA. 2016. Chemical proteomics approaches for identifying the cellular targets of natural products. Nat Prod Rep 33:681–708.

39. Jafari R, Almqvist H, Axelsson H, Ignatushchenko M, Lundbäck T, Nordlund P, Martinez Molina D. 2014. The cellular thermal shift assay for evaluating drug target interactions in cells. Nat Protoc 9:2100–2122.

40. Reinhard FBM, Eberhard D, Werner T, Franken H, Childs D, Doce C, Savitski MF, Huber W, Bantscheff M, Savitski MM, Drewes G. 2015. Thermal proteome profiling monitors ligand interactions with cellular membrane proteins. Nat Methods 12:1129–1131.

41. Kawatkar A, Schefter M, Hermansson N-O, Snijder A, Dekker N, Brown DG, Lundbäck T, Zhang AX, Castaldi MP. 2019. CETSA beyond Soluble Targets: a Broad Application to Multipass Transmembrane Proteins. ACS Chem Biol 14:1913–1920.

42. Ashok Y, Nanekar R, Jaakola VP. 2015. Defining thermostability of membrane proteins by western blotting. Protein Eng Des Sel 28:539–542.

43. Nji E, Chatzikyriakidou Y, Landreh M, Drew D. 2018. An engineered thermal-shift screen reveals specific lipid preferences of eukaryotic and prokaryotic membrane proteins. Nat Commun 9:4253.

44. Duvold T, Rohmer M. 1999. Synthesis of ribosylhopane, the putative biosynthetic precursor of bacterial triterpenoids of the hopane series. Tetrahedron 55:9847–9858.

45. Pan W, Zhang Y, Liang G, Vincent SP, Sinaÿ P. 2005. Concise syntheses of bacteriohopanetetrol and its glucosamine derivative. Chem Commun 3445–3447.

46. Piancatelli G, Luzzio FA. 2001. Pyridinium Chlorochromate, p. . *In* John Wiley & Sons, Ltd. (ed.), Encyclopedia of reagents for organic synthesis. John Wiley & Sons, Ltd, Chichester, UK.

47. Brown RT, Mayalarp SP, Watts J. 1997. Synthesis of methyl secolonitoside. J Chem Soc, Perkin Trans 1 1633–1638.

48. Higashibayashi S, Shinko K, Ishizu T, Hashimoto K, Shirahama H, Nakata M. 2000. Selective Deprotection of *t*-Butyldiphenylsilyl Ethers in the Presence of *t*-

Butyldimethylsilyl Ethers by Tetrabutylammonium Fluoride, Acetic Acid, and Water. Synlett 2000:1306–1308.

49. Wu C-H, Bialecka-Fornal M, Newman DK. 2015. Methylation at the C-2 position of hopanoids increases rigidity in native bacterial membranes. Elife 4.

www.ingramcontent.com/pod-product-compliance
Lightning Source LLC
LaVergne TN
LVHW042113190726
843493LV00006B/1467